BEYOND CLIMATE FIXES

From Public Controversy to System Change

Les Levidow

BRISTOL
UNIVERSITY
PRESS

First published in Great Britain in 2023 by

Bristol University Press
University of Bristol
1–9 Old Park Hill
Bristol
BS2 8BB
UK
t: +44 (0)117 374 6645
e: bup-info@bristol.ac.uk

Details of international sales and distribution partners are available at bristoluniversitypress.co.uk

© Bristol University Press 2023

British Library Cataloguing in Publication Data
A catalogue record for this book is available from the British Library

ISBN 978-1-5292-2238-8 hardcover
ISBN 978-1-5292-2239-5 paperback
ISBN 978-1-5292-2240-1 ePub
ISBN 978-1-5292-2241-8 ePdf

The right of Les Levidow to be identified as author of this work has been asserted by him in accordance with the Copyright, Designs and Patents Act 1988.

Cover design: Clifford Hayes
Front cover image: iStock/Sabrina Bracher

Contents

List of Figures and Tables

Figures

Tables

Glossary

Some of these concepts are elaborated in Chapter 1

Climate Justice   Climate Justice opposes socially unjust high-carbon political-economic arrangements, especially the global North imposing resource burdens and plunder in the global South, as well as injustices against frontline communities more generally.

Commoning   This is a process whereby a community maintains, manages and/or creates common resources. This effort also protects commons from enclosure by coercion or market competition (de Angelis, 2003, 2017; Wall, 2014; Bollier and Helfrich, 2015).

Co-production   This is a process constructing complementary forms of Nature, technoscientific knowledge and social order in a distinctive way (Jasanoff, 2004). For example, a co-production process may help construct a techno-market fix or else an eco-localization agenda.

Counter-publics   Counter-publics integrate diverse knowledges from socially marginalized groups and critical experts, facilitated by non-governmental organizations and social movements. Mobilized counter-publics generate public controversy over dominant agendas, undermine public consent for them, counterpose alternative futures and demand or create relevant knowledges for such aims (Hess, 2007, 2016; Frickel et al, 2010).

Eco-localization

Eco-localization builds more enjoyable lives by localizing production-consumption circuits that create low-carbon, resource-light, socially just livelihoods (North, 2010).

Ecological modernization (ecomodernism)

Ecological modernization seeks to overcome or avoid environmental degradation through technoscientific innovation that more efficiently uses natural resources, as a basis to reconcile environmental protection with economic growth.

Economic imaginaries

This is a societal vision whereby an 'imagined economic space' relates to an 'imagined community of economic interest' as a basis for articulating common strategies, projects and visions (Jessop, 2005). For example, the nation (or 'Europe') becomes a single competitive space facing a common external threat and market opportunity (Rosamond, 2002).

Food sovereignty

This is the right of peoples to healthy and culturally appropriate food produced through ecologically sound and sustainable methods, and their right to define their own food and agriculture systems (Nyéléni, 2007).

Grassroots innovation

This stimulates innovation processes that are socially inclusive towards communities as regards knowledges, processes and outcomes, thus empowering participants. These processes respond to social injustices, socio-economic inequalities and environmental problems often arising from conventional capital-intensive innovation (Smith et al, 2014; Levidow and Papaioannou, 2018).

Greenhouse gases (GHGs)

Greenhouse gases have two types. Carbon dioxide (CO_2) results mainly from combusting fossil fuels. Methane (CH4) results from leakages when extracting fossil fuels and routinely from agriculture (especially soil tillage and cows).

Neoliberal (or market) environmentalism	This largely relies on financial instruments as means to incentivize environmental improvement or protection. This policy framework generally weakens or limits statutory regulation (Bernstein, 2001; Bailey, 2007).
Sociotechnical	This refers to specific socio-economic relationships being embedded and facilitated within a technical design or process. These relationships are often disguised as merely technical instruments or as neutral characteristics (Law, 1986; Bijker, 1997).
Sociotechnical imaginary	These are 'collectively imagined forms of social life and social order reflected in the design and fulfilment of nation-specific scientific and/ or technological projects'. They articulate 'the relationship of science and technology to political institutions' (Jasanoff, 2004).
Techno-market imaginary	This is a societal vision whereby incentives for market competition stimulate techno-innovation for addressing societal problems, for example, alleviating resource burdens and climate change (Levy and Spicer, 2013). Such an imaginary combines ecological modernization with neoliberal environmentalism.

About the Author

Les Levidow is Senior Research Fellow at the Open University, UK. He has studied agri-food-environmental issues, especially technofixes, public controversy and alternative agendas. A long-time case study was controversy over agribiotech (transgenics) in the European Union, US and their trade conflicts. Other case studies have included controversies over biofuels, bioenergy and waste conversion. He has researched alternative agri-food systems and agroecology as a transformative agenda, initially European networks, and more recently South American ones for a solidarity economy and food sovereignty. He has been co-editor of the journal *Science as Culture* since the 1990s.

Acknowledgements

For advice on the overall book, I would like to thank the following people: academic colleagues (Davis Hess, Kelly Moore, Maria Nita, Luigi Pellizzoni, Brian Tokar), several friends (Anne Gray, Dave Rosenfeld, Simon Pirani) and the Bristol University Press STS Editor (Paul Stevens).

Thanks to colleagues on specific case studies:

Chapter 3, EU agribiotech: Susan Carr and David Wield (Open University) for support in several research projects, including teams in EU member states. Also Michel Pimbert (Coventry University) and others for the agroecology analysis.

Chapter 4, EU biofuels: Jenny Franco (Transnational Institute) and Helena Paul (EcoNexus) for joint research. Ariel Brunner (Birdlife International) for strategy insights.

Chapter 5, UK waste conversion: Paul Upham and Sujatha Raman for joint research. Shlomo Dowen (UKWIN) and Tania Inowlocki (XRZW) for strategy insights.

Chapter 6, Green New Deals: Adrienne Buller (Labour for a Green New Deal), Craig Dalzell (Common Weal), Stuart Graham (Glasgow TUC), Simon Pirani, Thea Riofrancos (Providence College) and Ellen Rowbottam (Leeds Trades Council) for strategy insights. Thanks to Leeds Trades Council for excerpts from its call for action to retrofit houses.

The four case-study chapters draw on some articles as follows:

Chapter 3

Levidow, L., Carr, S., Wield, D. (2005) EU regulation of agri-biotechnology: precautionary links between science, expertise and policy, *Science & Public Policy* 32(4): 261–276.

Levidow, L. (2009) Making Europe unsafe for agbiotech, in Paul Atkinson, Peter Glasner, Margaret Lock (eds), *The Handbook of Genetics & Society*, London: Routledge, pp 110–126.

Levidow, L., Pimbert, M., Vanloqueren, G. (2014) Agroecological research: conforming – or transforming the dominant agro-food regime?, *Agroecology and Sustainable Food Systems* 38(10): 1127–1155.

Chapter 4

Levidow, L. (2013) EU criteria for sustainable biofuels: Accounting for carbon, depoliticising plunder, *Geoforum* 44(1): 211–223.

Chapter 5

Levidow, L. and Upham, P. (2017) Linking the multi-level perspective with social representations theory: MSW gasifiers as a niche innovation reinforcing the energy-from-waste (EfW) regime, *Technology Forecasting and Social Change* 120: 1–13.

Chapter 6

Levidow, L. (2022) Green New Deals: what shapes Green and Deal?, *Capitalism Nature Socialism* (*CNS*) 33(3): 76–97, https://doi.org/10.1080/10455752.2022.2062675

1

Introduction to Climate Fixes versus System Change: What's the Problem?

The climate and ecological crisis cannot be solved without system change.
> Greta Thunberg, UN Climate Action Summit, 2020
> (EcoWatch, 2020)

'System change, not climate change' is not a request we make to the current institutions.
> Ecosocialist Encounter (2022)

Introduction

'System Change Not Climate Change' has become a more prominent slogan in recent years. It has sharpened public debate about the societal changes that are necessary to avoid climate disaster in ways creating an environmentally sustainable, socially just future. The demand for 'system change' directs attention at profit-driven, high-carbon production systems which cause climate change, other environmental harms, resource plunder and social injustices, along with policies which perpetuate them.

The slogan was promoted by the **Climate Justice** movement in the run-up to the 2009 Climate Summit in Copenhagen, where nearly 300 organizations launched a joint declaration for system change. Given the global elite's policy framework of carbon trading and carbon offsets, the signatories rejected such schemes as 'false solutions' (Klimaforum09, 2009). Back then, this confrontational approach was marginal within the wider climate movement.

Its mainstream organizations generally had accepted the elite's market-based policy framework, either reluctantly or enthusiastically. Representing the prevalent activism, the Climate Action Network (CAN) emphasized scientific imperatives for reducing **greenhouse gases** (GHG) reductions, along with fairness in the means. The latter meant the global North accepting more socially equitable targets and commitments within the UN Climate Convention (Hadden, 2015).

By contrast, the Climate Justice movement has linked diverse groups which see capitalism as a fundamental source of social injustice, climate change and wider environmental degradation. Climate Justice had antecedents and analogies with the 'environmental justice' movement departing from mainstream environmentalism. The latter attributes environmental degradation to specific actors (for example, companies, institutions or governments). By contrast, environmental justice perspectives attribute socio-environmental degradation to the neoliberal development paradigm, as the basis to confront complicit institutions (Šimunović et al, 2018). This diagnosis poses an imperative to transform production-consumption systems, moving from resistance to reconstruction (Schlosberg, 2013).

Extending that perspective, the Climate Justice movement opposed the official reliance on market-based measures to address climate change. It organized civil disobedience to confront the 2009 Climate Summit. It counterposed efforts to reclaim or create commons, that is, resources which are protected, managed and allocated by communities of various kinds (Chatterton et al, 2013).

That perspective had initially emerged at the first Climate Justice Summit a decade earlier. Affirming 'that climate change is a rights issue', it sought to 'build alliances across states and borders' against climate change for environmentally sustainable, socially just solutions (Karliner, 2000). The Climate Justice movement accused high-carbon industries of criminal irresponsibility. 'Climate Justice means opposing destruction wreaked by the Greenhouse Gangsters at every step of the production and distribution process.' Necessary remedies would include drastic domestic reductions in motor vehicle emissions, as well as more effective public transport, argued Corporate Watch (Bruno et al, 1999).

As another step, in 2004 the Durban Group for Climate Justice opposed carbon trading for 'privatizing the air', that is, selling the right to pollute the atmosphere. It anticipated that the European Union (EU)'s new Emissions Trading System (ETS) would not only fail to reduce net GHGs, but also would pre-empt more appropriate means to address the climate crisis (Bond, 2012: 31–32). In subsequent years these Climate Justice perspectives encompassed more land-based people's movements, many resisting dispossession (Tokar, 2013). Proponents of system change blame capitalist exploitation and free-market solutions for perpetuating several environmental and social injustices (Empson, 2019).

In those ways, the slogan has aligned issues, perspectives and social identities of diverse groups. Bridging their diverse frames has strengthened the Climate Justice movement (della Porta and Parks, 2017; Hadden, 2015; drawing on 'frame alignments' from Snow et al, 1986: 467). Although a confrontational approach initially separated the Climate Justice movement from mainstream climate groups, some eventually moved in a similar political direction. While the mainstream climate movement had emphasized 'science' as a key rationale for action, some groups shifted towards socio-environmental injustice.

The slogan became more prominent in the 2019 School Strike for Climate and then the Fridays for Future protests. There the demand for 'system change' gained relatively more support from women, non-White and lower-income participants (Todd Beer, 2022). Indeed, the slogan targets structural drivers of gender, racial and class oppression (della Porta and Portos, 2021). Likewise it indicates a broad social basis for a confrontational stance against the political-economic elites maintaining those structures (Figure 1.1).

For several decades a confrontational stance has come from frontline communities, sometimes called the Most Affected Peoples and Areas (MAPA). Within Extinction Rebellion (XR) the MAPA network has blamed global market coercion for driving economic growth which is socially and environmentally harmful: 'Basically, it's not just that fossil fuels are at the core of the economies, it's also that inequality on the planet is fossilised into the

Figure 1.1: Doctors for Extinction Rebellion

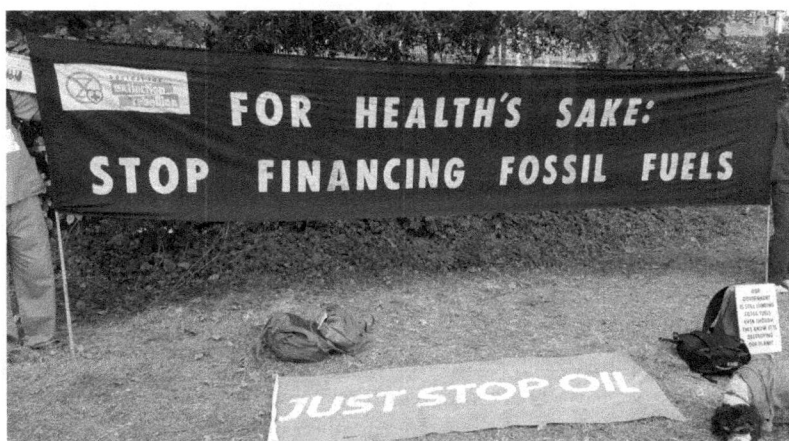

Note: 'Appreciating that Climate change is an impending public health catastrophe, we have decided to undertake civil disobedience with Extinction Rebellion.' This doctors' commitment illustrates the broader potential for multi-issue frame alignments within a Climate Justice perspective.

Source: Les Levidow

global financial structure.' Northern corporations exploit the odious national debt of Southern countries by negotiating contracts 'that allow the constant plundering of their territories through the export of commodities and extraction of fossil fuels, minerals, and other resources' (XR MAPA, 2022).

Greta Thunberg reprimanded a UN Climate Summit as follows:

> We are at the beginning of a mass extinction. And all you can talk about is money and fairy-tales of eternal economic growth. ... The climate and ecological crisis cannot be solved without system change. That's no longer an opinion. That's a fact. (EcoWatch, 2020; also Rowlatt, 2020)

A reprimand also came from a prominent official: 'Investing in new fossil fuels infrastructure is moral and economic madness', declared the UN Secretary General Antonio Guterres (2022b). Yet calling it madness obscures the prevalent economic rationale that has driven resource plunder for many centuries. From a Climate Justice perspective, European colonialism 'set in motion a global model for racialised resource extraction from people of colour', especially in the global South; its debt repayments further drive the plunder today (Greenpeace and Runnymede Trust, 2022: 12, 17).

Global elites have perpetuated high-carbon systems in overt ways, especially by expanding fossil fuels. They have also encouraged techno-optimistic promises for low-carbon solutions, promised smooth pathways to decarbonization, encouraged a passive public to accept or await such fixes and thus depoliticized or pre-empted societal choices about potential futures. A key instrument has been market-type incentives, often serving to perpetuate high-carbon systems. This techno-market framework has been central to the UN Climate Convention and the EU, despite their pretensions to global environmental leadership.

Such political evasions take various forms. Some proponents have idealized future technologies as 'climate fixes' which would avoid the need for major societal change and so be more feasibly implemented (Pielke, 2010). To reach the target of near-zero emissions, 'I am told by scientists that 50% of the reductions we have to make by 2050 are going to come from technologies we don't yet have', said the US government's climate envoy John Kerry (quoted in Harrabin, 2021). His comment was widely ridiculed as a faux pas or naive wish. Yet his wishful expectation revealed the elite's long-term alibi, namely: awaiting hypothetical fixes, perhaps even funding them, meanwhile continuing high-carbon production and consumption.

Over the past decade or so, more environmentalist groups have denounced such fixes as 'false solutions' which evade the systemic causes of climate change and thus perpetuate them. Critics have drawn analogies with other environmental problems and their putative technofixes as false solutions. Such controversy has opened up greater opportunities to promote low-carbon, lower-energy alternatives.

Lacking an effective social agency, however, such proposals often have remained as technical-administrative blueprints, or as futile appeals to hypothetical planners. Even when such plans are implemented, the solutions often become low-carbon supplements to high-carbon systems rather than replacements. To gain at least some socially just decarbonization measures, the climate movement necessarily makes demands on the dominant institutions. However, as one activist network argues:

> 'System change, not climate change' is not a request we make to the current institutions. It is our responsibility to make it happen. To achieve this requires that we be coordinated globally and regionally, that we define strategies and act together, and create spaces where we can build peoples' power and grow the movement. (Ecosocialist Encounter, 2022: 11)

How, then, to create the necessary social agency? How to replace harmful production systems, moving beyond climate fixes towards system change? Such efforts will be illustrated here by several case studies and then analysed through cross-case comparisons. Together these analyses provide the means to identify such efforts more widely and to strengthen them. But first this introductory chapter will examine a general obstacle in techno-market fixes, as a basis for devising strategies against and beyond them.

Climate fixes have facilitated high-carbon continuity

Protest against incumbent high-carbon systems has provided new imperatives and opportunities for substitutes which can significantly reduce GHG emissions, alongside a general reduction in energy usage, especially in the global North. Yet promises of climate fixes, and sometimes their implementation, have often facilitated high-carbon continuity. Political-economic elites have evaded or rejected the necessary means to decarbonize economies, given at least three political-economic constraints.

As one constraint, full decarbonization would mean abandoning the great 'sunk investment' in fossil fuels and the wider production systems dependent on them (Carton, 2019). Although more expensive than water mills, fossil fuels originally gained predominance for several political-economic reasons, especially their profitability, as well their flexibility for disciplining human labour; such capitalist advantages persist today (Malm, 2016). Global elites have sought various means to protect their past investment as capital assets (Geels, 2014). For many years, environmental economists have warned about the prospect of 'stranded assets', that is, fossil fuel deposits whose market value may decline. Nevertheless economic elites have sought ways to protect their high-carbon financial assets, even if simultaneously diversifying into renewable energy.

As a second constraint, only approximately one-fifth of final energy use is electricity and thus readily amenable to substitution by renewable sources. Of the total electricity supply during 2017–2019, only a quarter came from renewable sources, with only a modest increase (IEA, 2021a). Renewable energy prices may decline further, relative to fossil fuels; yet the latter offer a more profitable return on investment, at least under prevalent government policies (Christophers, 2022). Processes directly using fossil fuels would need to be electrified in order for renewable energy to become a commercially feasible substitute. More fundamentally, there would need to be a significant decline in energy use, leading to the next point.

As a third constraint, comprehensive decarbonization would need structural changes beyond market-type incentives and competition. Public-sector institutions would need to take political responsibility and make enormous direct investments alongside civil society mobilizations, towards transforming economic sectors. This imperative 'changes everything' and so has provoked elite denials or evasions of the problem, as analysed by Naomi Klein (2014).

Echoing her diagnosis, Jonathan Neale observes:

> Anyone who thinks about climate solutions for any length of time realizes it will take massive government action; it will have to move far beyond the rules of the market; and it will change everything. The people who now run and own the world have spent their adult lives convincing the rest of us of three things. First, we have to obey the rules of the market. Second, there is no alternative to the market. Third, if we disobey, we will be crushed. The most powerful weapon the rich and powerful have is that we believe them. … So the majority of the rich and powerful want to do what needs to be done, but they cannot. (Neale, 2021: 210)

Given its incapacity or refusal to make such changes, the global elite reinforces and extends market-type rules as if they were natural. These drive various fixes, which have two elements: promises of future decarbonization technologies, alongside market-type incentives for such innovation. Let us briefly survey their complementary roles.

As the most prominent example, Carbon Capture and Storage (CCS) has carried the future promise to avoid GHG emissions from fossil fuels, especially as a basis to continue such fuels in state scenarios of low-carbon futures alongside economic growth. CCS plants also may be eligible to provide carbon credits for sale. With these promises, promoters have justified state funds for CCS research and development (R&D). According to early critics, such investment 'squanders finance on transferring fossil fuels out of the ground while delaying transitions to non-fossil technologies' (Lohmann, 2009).

As another wishful fix, geoengineering was originally promoted as a last resort if GHG emissions accelerate despite efforts to reduce them. 'Therefore, the last best hope may reside in an environmental fix, engineered independently of energy systems transformation, namely radiation management that cools down the planet', said a prominent scientist (Schellnhuber, 2011: 20277). As this illustrates, climate 'fixes' can have a favourable meaning. As promoted by some experts, technological solutions can gain popular appeal as a reassuringly smooth means to address environmental problems.

However, the term 'fix' has become generally pejorative, as in the sarcastic phrase 'technofixation'. Climate fixes have been criticized on various grounds, for example, as illusory, unreliable, technically unfeasible, burden-shifting and/or potentially unjust: 'The persistent claim that a solution is just around the corner has allowed politicians and corporations to cling to the mantra that tackling climate change will not impact on economic growth. ... Technofixes appeal, in short, to the powerful, because they offer an opportunity to maintain power and privilege' (Corporate Watch, 2008: 1, 7). Such solutions evade the causes of climate change, perpetuate fossil fuels and may introduce new risks (ETC Group, 2015; FoEE, 2015).

Future technological promises to remove carbon underlie the 'net-zero carbon' agenda, which has been emerging as a joint state–business deal for green transition (Beuret, 2021). Although carbon removal initially attracted many climate scientists, some have criticized its dependence on an elusive technological salvation. It 'has licensed a recklessly cavalier "burn now, pay later" approach which has seen carbon emissions continue to soar' (Dyke et al, 2021).

Policy elites and some non-governmental organizations (NGOs) have celebrated or urged increases in renewable energy. Although necessary, this has largely supplemented fossil fuels rather than replaced them. Before the COVID-19 pandemic disrupted economic activity, renewable sources provided only one-quarter of the global growth in total primary energy usage; the latter increased twice as fast as economic growth, especially through natural gas (IEA, 2019). In 2020 nearly all fuel-supply investment went into fossil fuels: 84 per cent to oil and gas and 14.5 per cent to coal; only 1.3 per cent went to low-carbon fuels, despite their declining cost (IEA, 2021b). In 2021 renewables investment rose, but energy usage rose even more, mostly coming from the extra production of coal and gas (IEA, 2022).

The UN Secretary General has denounced the rise: 'This abdication of leadership is criminal.' He counterposed renewables as the remedy: 'It is time to stop burning our planet and start investing in the abundant renewable energy all around us' (Guterres, 2022a, 2022b). Yet renewables expansion obscures the parallel rise of energy usage and fossil fuels.

In the global North, economies have been somewhat decarbonized mainly by deindustrializing them and/or replacing coal with natural gas. Their GHG emissions have been outsourced to the global South, for example, through high-carbon imports or offsets via carbon trading. The latter fix illustrates techno-market agendas, which are explained next.

Market-type instruments have blurred responsibility

As a key driver of climate change, resource-intensive, high-carbon projects have been expanded globally, thus undermining the decarbonization targets of global climate agreements. But those agreements are not innocent. Let us examine their political roles in system continuity through techno-market fixes. This framework promotes market-type incentives in order to stimulate eco-efficient technoscientific solutions; their future promises (even failures) have justified more market incentives, through a recurrent circular reasoning.

Techno-market fixes eventually emerged from the 1992 UN Framework Convention on Climate Change (UNFCCC), henceforth the UN Climate Convention for short. Its implementation depended on the 1997 Kyoto Protocol, entering force in 2005. According to early drafts, GHG emissions would have global and national caps, to be gradually reduced, thus stimulating steady reductions in GHGs. This proposal was known as 'cap and trade'.

However, the cap was ultimately omitted. State-corporate lobbies advocated tradeable carbon credits: polluters could offset their emissions by buying surplus credits. The US chief negotiator, Vice-President Al Gore, imposed this arrangement as a condition for US government support – which anyway never resulted in US ratification. On this empty promise, the 1997 Kyoto Protocol mandated carbon trading schemes, meant to reward and stimulate low-carbon technologies; hence it exemplifies a techno-market framework.

Carbon trading schemes have generally allocated emissions permits to carbon polluters on the basis of recent emissions. These schemes have effectively subsidized fossil fuel companies, with doubtful GHG reductions. As an original rationale for this techno-market fix, the rising carbon price would incentivize a shift to low-carbon technologies. In practice, especially given a low carbon price, carbon trading schemes have licensed companies to continue their carbon pollution, while also reinforcing their political power. This policy framework has helped elites to avoid stringent environmental regulation and significant decarbonization. For those reasons, activists have demanded that UN climate talks exclude 'climate criminals' and 'polluters' who avoid responsibility through carbon offsets (Cintron, 2015).

What has this meant for low-carbon substitution? As a prominent scheme, the UN's Clean Development Mechanism (CDM) has awarded Certified Emissions Reductions (CERs) on the basis that a development project will generate less GHG emissions than would have happened otherwise. CDM proponents have anticipated this counter-factual scenario as pessimistically high carbon (for example, coal or oil), so that they can portray their technological option as comparatively low carbon. In many cases these development projects have marginalized local communities, restricted their access to natural resources, undermined their livelihoods and/or pre-empted their development agendas. In practice the CDM has shifted political-economic power to those who buy carbon credits from such 'development' projects in the global South (CMW, 2018).

This techno-market framework became the basis for negotiating rules at the UNFCCC's regular Conferences of Parties (COP). Carbon-credit trading was reinforced in the 2012 Paris Agreement. Its Article 6 established a 'market-based mechanism' structuring 'the use of internationally transferred mitigation outcomes', that is, carbon trading to offset emissions in the global North (UNFCCC, 2012). This policy framework has become increasingly blamed for undermining peasants' rights to land and natural resources in places that generate the carbon credits (GRAIN and WRM, 2015).

At the UNFCCC's COPs, the agenda has recurrently focused on criteria for carbon credits, offsets and markets. At the 2019 Madrid COP25, this focus marginalized discussion of social and environmental safeguards (Goldberg, 2019). At the 2021 Glasgow COP26, the agenda put great effort into finalizing the criteria, to make such schemes fully operational. 'Business are keen to use offsets to garnish their green credentials or as an opportunity to make it cheaper to reach their targets', reported Bloomberg News (Krukowska, 2021).

Through its financial instruments, the techno-market framework displaces political responsibility to anonymous market forces and so depoliticizes societal choices. As a critic argued early on, 'Many governments are meanwhile hoping that major climate investment decisions can be simply left to the new carbon markets' (Lohmann, 2009: 1064; see update, Lohmann, 2020). In emissions trading schemes, agents' decisions are shaped by market prices rather than by governmental agencies (Page, 2012). In practice such schemes have prioritized the cheapest means of apparent GHG reductions, which are often doubtful or illusory, while avoiding responsibility for more expensive reductions. This arrangement has perpetuated production systems which extract and burn fossil fuels (Clare, 2019). This general role warrants scrutiny of the EU's climate pretensions, as analysed next.

EU climate policy has depended on techno-market fixes

The EU has gained a reputation for the most stringent environmental standards and thus global leadership. Although valid for some environmental issues, this reputation is contradicted by its climate policies. The EU's neoliberal regime has helped perpetuate harmful production systems, as shown here.

The EU ETS was the world's first scheme seeking to incentivize low-carbon technology under the Kyoto Protocol: 'Trading brings flexibility that ensures emissions are cut where it costs least to do so. A robust carbon price also promotes investment in clean, low-carbon technologies' (EU ETS, 2005; also EU ETS, 2015: 14). Like the Kyoto Protocol, the EU ETS depended on optimistic assumptions, especially that a high carbon price would incentivize low-carbon substitutes.

Many environmental NGOs welcomed the EU ETS, especially its promise to assist decarbonization measures in the global South (for example, EDF, 2007: 2). Nevertheless some NGOs highlighted problems, especially for the global South. The EU ETS had failed 'to deliver its fair share of the finances and technology needed by developing countries to tackle climate change' (FoEE, 2010a: 10).

The EU ETS instead provided 'supplementary finance for fossil-intensive industrial pathways in the South' (Lohmann, 2009). It accelerated a shift from coal to natural gas rather than to renewables (Reyes, 2012). Moreover, it has served political agendas for avoiding regulatory measures that would mandate significant GHG reductions (Reyes, 2014).

A decade after the EU ETS began, its failure became more obvious. It had not significantly driven Europe's GHG reductions; these were due mainly to other factors, especially its economies deindustrializing and outsourcing emissions. Meanwhile the ETS was undermining alternative decarbonization measures, such as a cap on carbon emissions, while even subsidizing polluters (CEO, 2015). 'A weak reduction target and the massive use of international offsets have led to the build-up of and enormous surplus of emission allowances. ... This risks a lock-in of carbon intensive infrastructure for years to come', warned the CAN, representing mainstream NGOs (CAN, 2017).

Despite those warnings, the ETS has extended its earlier technological promises and market-type measures. For the ETS's third phase during 2013–2020, the New Entrants' Reserve raised funds from carbon-credit sales in order to fund various technologies. These included: CCS (for example, pre-combustion post-combustion, oxyfuel and industrial applications) and renewable energy, especially bioenergy, solar power, photovoltaics, geothermal, wind, ocean, hydropower and smart grids (CEC-Innovation,

2019). Such promises again wishfully conflated technoscientific innovation with low-carbon substitution of high-carbon systems. This prospect has been weakened or delayed by dependence on carbon trading, thus pre-empting more reliable means such as a cap on carbon emissions.

Techno-market fixes are further exemplified by the 2019 European Green Deal. This was meant to integrate GHG reductions into the EU's economic policy so that technological innovation achieves a zero-carbon Europe by 2050. 'Our goal is to reconcile the economy with our planet, to reconcile the way we produce and the way we consume with our planet and to make it work for our people', declared the European Commission President (von der Leyen, 2019). How?

As promised in the EU's earlier policy, public subsidy would establish market incentives nudging private-sector finance. The EU would become 'a global leader' against climate change by innovating low-carbon, eco-efficient technologies and exporting them. The EU's R&D programme will 'leverage national public and private investment' to 'facilitate the phasing out of fossil fuels, in particular those that are most polluting, ensuring a level-playing field in the internal market'. Beyond fair competition within the EU, it will further 'develop international carbon markets as a key tool to create economic incentives for climate action', according to the European Green Deal (CEC, 2019: 18, 20).

Its agenda has relied on technological innovation, portrayed as politically neutral: 'it is essential to ensure that the European energy market is fully integrated, interconnected and digitalised, while respecting technological neutrality'. It advocates 'smart infrastructure', in particular: 'innovative technologies and infrastructure, such as smart grids, hydrogen networks or carbon capture, storage and utilisation, energy storage, also enabling sector integration' (CEC, 2019: 6).

Technological development to capture carbon from natural gas for 'blue hydrogen' has been arguably less about decarbonizing the economy than about capturing any future hydrogen market, thus further capitalizing the infrastructure and sunk assets of fossil energy. This future pathway has been favoured by the EU's regulatory criteria (CEO, 2020; Szabo, 2021). Under its wider policy framework, any decarbonization failure would be due to faulty market incentives or technological limitations, thus displacing and evading the state's responsibility.

Moreover the EU's decarbonization policy has shifted seed funding towards private-sector investment. It depends on the European Investment Bank (EIB) offering public guarantees to attract private co-investments. By contrast, the EU's earlier development schemes had involved partnerships with regional and local authorities. They were democratically accountable voices, potentially advocating different investment priorities, but later became marginalized (Neujeffski, 2021).

For such reasons, the European Green Deal worsened the EU's democratic deficit. According to a civil society alliance:

> The so-called 'Sustainable Europe Investment Plan' does not provide resources for communities, municipalities, or regions to invest in their housing or utilities. Instead, it subsidises private investors, socialising the risks of the green transition while privatising the gains. Those who live in Europe are given no control over the direction of Europe's decarbonisation. (GNDE, 2019: 20)

This anti-democratic arrangement results from measures which embrace markets as the social order: 'They follow the same market-based, technocratic and instrumental rationale of the ecological modernisation paradigm. That is, they *change the instruments* of production, consumption and growth, but *do not question the goals* of endless production, consumption and growth' (Huber, 2020: 4, emphasis in original).

In the EU's mid-2021 climate agenda, its carbon trading scheme was extended to transport systems and buildings' heating systems, featuring auctions of carbon credits. The scheme was meant to stimulate low-carbon techno-innovation. It was criticized for incapacity to deliver a low-carbon transition, or for unjust burdens on lower-income people, or both (Taylor, 2021).

Why this pattern of false promises and social injustice? The EU's beneficent green narrative promotes and disguises a market-type policy framework as a decarbonization agenda. Alongside the green facade, it extends previous high-carbon commitments and investments. Meanwhile the EU's austerity rules often constrain state investment in truly renewable energy. The EU greenwashes its policies by co-opting language from the climate movement (Varoufakis and Adler, 2020). Within this policy framework, no one is assigned ultimate responsibility for decarbonization failures. This leads to the next point about social agency remaining elusive in proposals for climate solutions.

Social agency remains elusive for climate solutions

In the burgeoning literature on climate change, various solutions differ on many fundamental issues such as economic growth, market drivers and innovation. Alongside such differences, most proposals leave ambiguous the relevant social agency that could implement solutions such as decarbonizing or replacing high-carbon systems. Here the term 'social agency' includes the political will, collective capacities and necessary resources to impel or implement solutions. Social agency is often left elusive by verbs in the passive case, or by active verbs referring to market-type incentives.

An important social agency has come from frontline communities, meaning those groups most experiencing harms from climate change and high-carbon systems (for example, XR MAPA, 2022). Such groups are generally low-income, racially dispossessed groups which lack basic infrastructure to support them and who will be increasingly vulnerable to an erratic climate. Frontline communities have been opposing harmful high-carbon developments and false solutions for a long time, while extending traditional resource-light production methods as alternatives.

Frontline communities face neocolonial, technocratic policies and greenwashed alternative energy infrastructures throughout the global South. Through resistance struggles and alternatives, social movements seek to transform their habitats for a dignified survival (Gelderloos, 2022). Such practices have inspired many activists worldwide. But strategies there are less clearly transferable to institutional contexts and political forces in the global North, which warrants its own analysis.

As regards the global North, one author proposes that its future economic growth should be restricted to zero-carbon means, while shifting consumption from an individual basis to group or public services. For the UK as an example, his book provides administrative-technical blueprints for decarbonizing each economic sector through changes in materials and energy usage (Neale, 2021). Through such blueprints, systemic change can seem more credible – except for lacking a clear social agency. Another diagnosis identifies the neoliberal shifts and harmful economic growth in recent decades as a main obstacle, to be reversed and overcome through a degrowth strategy. Jason Hickel advocates several pathways, such as: to decommodify public goods, expand the commons, shift consumption from individual ownership to group usership, and shift employment to socially useful tasks including social reproduction (Hickel, 2020).

Such schemas and principles can stimulate debate on alternative futures. Yet the pathways have a general lacuna about social agency. The proposals may assume or imply that people first must be persuaded about specific futures before being mobilized. In practice the reverse often happens: when activists try to block a harmful development or to counterpose a low-carbon alternative, they bring diverse priorities and future societal visions. This process can clarify objectives and stimulate political alliances to realize them.

For social agency in the global North, various alliances have been campaigning for structural change linking social justice with decarbonization. Such alliances have often adopted the banner of a Just Transition or a Green New Deal (GND). From a perspective on the climate crisis as class struggle, trade-union power must decarbonize high-carbon sectors such as energy, food, housing and transportation, argues Matt Huber (2022). Local climate plans 'can too easily simply provide another opportunity for corporate interests to dominate'; therefore trade-union representation would be necessary to ensure wider societal benefits (CACCTU, 2021: 93).

Yet trade unions have divergent stances on issues such as decarbonization pathways, technological solutions and economic growth (TUED, 2015, 2019; see Chapter 6). The fossil fuel industry and the sector's trade unions have jointly promoted a CCS fix which would perpetuate fossil fuels. How to disrupt this cross-class partnership?

Towards a social agency for system change, deceptive climate fixes have sometimes provoked protest. This has brought together people with disparate entry points and understandings. Although they may remain fragmented, sometimes they have coalesced around low-carbon, socially just alternatives. How to understand and strengthen such mobilization strategies? How to draw lessons for future efforts? These questions lead to the big picture in this book.

Big picture: book overview

As this Introduction argues, techno-market fixes have pervaded dominant policy frameworks on climate change. In recent decades a similar pattern has arisen around other environmental issues, whose fixes likewise have provoked protest, controversy and alternatives. What can be learned from those experiences to inspire and guide future efforts?

That question can be best answered by political alliances testing strategies in practice, learning from experience and then modifying strategies. To inform such efforts, this book provides a big picture: it integrates several analytical concepts and then applies them to case studies (Chapters 3–6). These provide a basis to identify general patterns and draw some lessons. The following subsections present chapter summaries.

Chapter 2 on techno-market fixes: key concepts

Technofixes for environmental problems have often provoked controversies. As a pervasive policy framework in recent decades, techno-market fixes arose from a convergence between **ecological modernization** and **neoliberal environmentalism**. This policy framework has been elaborated by the UN Climate Convention and EU for a long time. The concept 'market' has gained multiple meanings, for example, as financial incentives, competitive imperatives, eco-efficient solutions and economic benefits. Although these fixes may be accepted by passive publics, **counter-publics** have linked citizens' groups with critical experts, thus generating greater public controversy. This has broadened opportunities for various alternatives which incur lighter resource burdens, enhance socio-economic equity, localize production-consumption circuits and devise technical means for those aims. Such alternatives become more than mere technical-administrative blueprints when multi-actor alliances become counter-publics confronting the dominant agenda; their alternatives often combine **eco-localization** with **grassroots innovation**. Rival agendas can be understood as **sociotechnical**; each agenda expresses a **sociotechnical imaginary** co-producing a distinctive form of natural resources, technoscientific knowledge and social order. Together these concepts provide a big picture relevant to various fixes, their controversies and alternative futures.

Chapter 3 on the EU agribiotech fix

The EU originally promoted agribiotech (genetically modified [GM] crops) through several policy changes extending market relations. This techno-market framework included broader patent rights, market liberalization of agriculture, and research agendas blurring the public and private sectors. In the EU's narrative, agribiotech would be a crucial eco-efficient means for multiple benefits, for example, for the sector to gain global economic competitiveness, to minimize farmers' dependence on agrichemicals and thus to protect natural resources. But GM crops were soon denounced for threatening the environment and human health. Mass opposition eventually blocked a European market, opening up opportunities for 'quality' alternatives, eventually for promoting agroecological systems. Yet policy support measures have been constrained by the dominant techno-market agenda, subsidizing agri-industrial systems for higher yield and global markets. In 2014–2015 GM crops were relaunched for a 'climate-smart agriculture' which supposedly would make agri-industrial systems more climate-resilient, while becoming eligible for carbon credits. Critics turned this agenda into a political controversy over 'corporate-smart greenwash' and thus an opportunity to promote agroecological alternatives as truly climate-resilient by 'cooling the planet and feeding the people'.

Chapter 4 on the EU biofuels fix

In 2009 the EU's Renewable Energy Directive set a statutory mandate for renewable energy in transport fuel, citing the need to reduce GHG emissions. This mandate expanded biofuels from edible feedstock, thus provoking controversy over multiple harms. The obligatory market incentivized land-use changes in the global South, in turn restricting resource access for local food production, while disguising various harms through sustainability criteria. As a key rationale, the EU mandate would stimulate the EU's Knowledge-Based Bio-Economy (KBBE): technoscientific progress would eventually bring 'advanced biofuels' using only non-edible feedstock, enhancing environmental sustainability. Meanwhile this techno-market framework served to perpetuate conventional biofuels and the internal combustion engine. Even before the Directive's enactment, critics provoked opposition over several issues, for example, the EU's resource plunder, a 'carbon time bomb', significant harms beyond the sustainability criteria, lax standards for vehicle emissions, and a delay in electric vehicles as replacements. Despite such broad opposition, the mandate continued to permit the most harmful feedstock for at least another decade after the 2009 Directive, thus perpetuating system continuity. This outcome resulted from prioritizing an 'investment climate' for the KBBE rather than GHG reductions.

Chapter 5 on the UK's waste-conversion fix

The EU's waste-management framework has been implemented through a UK techno-market framework shaping waste-energy trajectories. To reduce disposal of Municipal Solid Waste (MSW) via landfill, local waste authorities have outsourced responsibility through competitive bidding for waste-conversion plants, mainly incinerators. These plants provoked much controversy over local environmental nuisance, as well as a long-term waste generation 'to feed the beast'. Market-based incentives stimulated a techno-innovation, Advanced Thermal Treatments (ATTs); these were meant to avoid those problems, better recover or reuse waste, and thus save GHG emissions. Yet this techno-market fix perpetuated the previous problems; it also incentivized conversion processes which were closer to mere disposal, mainly as feedstock for distant energy production. Opponents have denounced all incineration for wasting resources, worsening the climate problem, imposing environmental injustices and perpetuating the linear economy. They counterposed a circular economy that would redesign production systems to reduce or reuse waste and to address fuel poverty; this provides a future vision for local alternatives and alliances promoting them.

Chapter 6 on Green New Deal agendas

In recent years, GND agendas have gained significant support for a transition to an environmentally sustainable, low-carbon, socially fairer economy. In the 2019 US and UK versions, endorsed by some public-sector trade unions, the GND sought to achieve a net-zero carbon by 2030 and reduce resource burdens. Advocates envisaged greater socio-economic equity by means such as expanding public goods, building workers' cooperatives and localizing production-consumption circuits. When these GND agendas sought endorsement by major political parties, however, trade unions in fossil fuel sectors sought a commitment to CCS as a condition for their support, thus perpetuating fossil fuels. Such false promises illustrate the general appeal of technofixes to soften societal conflicts around a potentially disruptive decarbonization process. By contrast, going beyond climate fixes, labour–movement alliances have been promoting GND local campaigns for a socially just, low-carbon economy. Their agendas for retrofitting houses illustrate a cooperative eco-localization perspective, in conflict with the neoliberal techno-market fix of competitive tendering with its minimal standards.

Chapter 7: Conclusion

Techno-market fixes carry beneficent promises to decarbonize economies in ways avoiding societal disruption and conflict. Although these promises have been rightly criticized as false solutions, the problem runs more deeply: they justify institutional change along neoliberal anti-democratic lines, supposedly in order to realize the techno-promises. Often a mobilized counter-public has stimulated public controversy and promoted alternative solutions. As many cases here illustrate, technical designs and standards always facilitate one social order rather than other, thus warranting political struggle. Exemplifying eco-localization, some local agendas would incur lighter resource burdens, enhance socio-economic equity, localize production-consumption circuits, assign political responsibility, encourage grassroots innovation and thus devise appropriate sociotechnical means. Such transformative mobilizations undermine climate fixes and go beyond them. More effective strategies can emerge from Participatory Action Research (PAR), whereby researchers and practitioners jointly define the societal problems that warrant research. In the case studies here, knowledge exchange with political activists helped to sharpen action-research for a systemic perspective on false solutions versus alternatives. This big picture can help to identify and facilitate an effective social agency for transformative mobilizations, as steps towards system change.

2

Techno-market Fixes Provoke Controversies and Alternatives: The Big Picture

Over the past half-century, technological solutions have been anticipated or promised for many environmental problems which often resulted from previous technologies (Rosner, 2004). The recurrent promise has been satirized as follows: 'The deterioration of the environment produced by technology is a technological problem for which technology has found, is finding, and will continue to find solutions.' Within this circular logic, in practice, 'technological innovation and efficiency improvements will continue to promote unsustainable growth' (Huesemann and Huesemann, 2011: 77, 116). Indeed, invoking greater efficiency, technological advance has often incentivized economic growth and larger markets, thus aggravating resource burdens and environmental degradation.

In some cases, a techno-innovation has been ridiculed as 'a solution in search of a problem'. Hence the well-known aphorism: 'I suppose it is tempting, if the only tool you have is a hammer, to treat everything as if it were a nail' (Maslow, 1966: 15). Originally called 'the law of the instrument' in psychology, this aphorism highlights how technical tools narrowly define a problem, thus implying a simple solution.

Given the numerous potential hammer-fixes, some have gained relatively more elite support and influenced societal futures. Which fixes, why and how? In recent decades, EU policy has emphasized techno-market fixes. To investigate them, here are some questions:

- How do policy frameworks promote market-type incentives and competition, supposedly in order to generate technological fixes for environmental problems (especially climate change)?
- How do those fixes encourage a passive public to accept or await them, meanwhile continuing harmful production systems?

- How do opponents contest those fixes, stimulate public controversy and so open up different societal futures?
- How do such opponents attempt to build a social agency for alternative solutions?
- In those ways, how do claims for solutions promote divergent societal futures, serving either system change or continuity?

Beyond each case: How can cross-case comparisons inform strategies for system change?

To help answer those questions, this chapter first explains some analytical concepts, in particular: rival social orders, techno-market fixes, sociotechnical imaginaries, transformative mobilizations and PAR. Together these concepts provide a big picture to illuminate the case studies here, as well as other controversial cases.

Rival social orders around technofixes

Some technological solutions have gained a broad appeal by apparently avoiding the need for systemic changes which seem politically awkward, economically disruptive or both. More than simply an error, '[t]he preference for solving problems with technology is driven by deep-rooted habits of thinking and tremendous institutional momentum within public and private research', argues the sociologist Noel Scott (2011: 208).

Such fixes can temporarily appear to address environmental problems. They have elicited popular ambivalence, combining both wishful and sceptical attitudes. As Leo Marx (1983) has argued, technofixes are embedded in our culture's dominant concept of history, as if societal change always follows from technological change and thus depends on such change for future progress. Within that dominant concept, policy frameworks obscure or disparage alternatives which have already provided solutions – and which could do so even more.

Consequently, environmental technofixes have become controversial for a long time. Although some criticisms focus on 'unintended side-effects', these may be in-built outcomes of a social model and sociotechnical design. At first sight, such fixes may substitute for societal change, albeit inadequately for effective solutions. But fixes embed policy assumptions about the necessary or desirable social order, thus potentially bringing societal change, but of what kind? In technofix controversies, rival visions of the most useful environment have provoked conflict as surely as rival visions of the optimal government (Rosner, 2004). This role warrants a closer look at relationships between technical, social and institutional change.

A half-century ago the term 'technological fix' was popularized by the former director of Oak Ridge National Laboratory, Alvin Weinberg. Previously he had celebrated nuclear power as means to avoid societal conflicts over resources. He posed this semi-rhetorical question: 'Can we identify quick technological fixes for profound and almost infinitely complicated social problems?' Such solutions would be more feasible than 'social engineering' of human behaviour, given its complex and unpredictable character. Hence engineers could be the best problem-solvers, he argued (Weinberg, 1966: 4; Johnston, 2018). He promoted cheap shortcuts 'that are within the grasp of modern technology, and which would either eliminate the original social problem without requiring a change in the individual's social attitudes, or would so alter the problem as to make its resolution more feasible' (Weinberg, 1967: 2, 9). As he later reflected, 'technological fix' connotes 'technical innovations that could help resolve predominantly social problems' (Weinberg, 1994).

Indeed, technological innovation has been recurrently sought in order to avoid societal change: 'Calling for innovation is, paradoxically, a common way of avoiding change when change is not wanted. The argument that future science and technology will deal with global warming is an instance. It is implicitly arguing that, in today's world, only what we have is possible' (Edgerton, 2006: 210).

Even when collective demands arise for significant change, often it has been avoided through elite techno-promises of future benefits. This response has protected previous industrial investments and the dominant political-economic power. For such reasons, technofix critics have emphasized the need for social and institutional change beyond or alongside technological change. Science-led technological innovation has a limited capacity to fix societal problems, which instead warrant 'social policy' (Sarewitz, 1996).

Each in its own way, these arguments imply that technological and societal change are separable. But are they? On the contrary, any technological innovation has a social character, more amenable to ordering society in one way, for example, by facilitating either cooperative or competitive relationships. From its earliest design, innovation is always sociotechnical. It emerges from sociotechnical networks of diverse actors: each brings their own expertise, frames, problem-definitions and evaluation criteria, which may be reconciled within a single sociotechnical framework and design (Bijker, 1997; see also Law, 1986). While the concept 'sociotechnical network' can explain multi-actor cooperation, it can likewise explain conflict, whereby different agendas promote rival models of sociotechnical change.

In early 2020, for example, the COVID-19 pandemic stimulated debate over how to limit the spread. Some experts posed a binary choice: either greater state techno-surveillance could exercise control over public behaviour, or else individual consumers could purchase data-tracing apps to protect their privacy. Beyond those contentious options, a different future could favour public sovereignty over digital platforms as a public good for mutual aid and solidarity, as advocated by Evgeny Morozov (2020). As he had earlier argued more generally, 'technological solutionism' presumes a specific problem with a simple solution. In a satirically idealized version, all our problems can be solved by the correct combination of computer codes, algorithms and robots (Morozov, 2013). Prevalent frameworks promote and disguise a specific sociotechnical order as if it were merely technical. In response, public debate can highlight and contest that implicit social order.

Social order likewise arises in various climate fixes known as climate engineering. These include techniques to remove carbon dioxide from the atmosphere, or to reflect solar energy back to space and thus rapidly cool the Earth. Beyond concerns about environmental risks, these technological trajectories have undergone diverse normative criticisms. On the one hand, given the technology's promise of a 'last resort' solution, this option may weaken efforts to slow down climate change and to mitigate its harmful effects. On the other hand, climate engineering technologies may justify a specific socio-political change (Kreuter, 2015: 21). In particular, they could favour narrow knowledge networks, 'highly dependent on top-down expertise and with little space for dissident science'; they would even depend on 'autocratic governance', thus undermining democracy (Szerszynski et al., 2013: 2812).

Regardless of such sociotechnical models, state agencies have invested scant funds in climate engineering technologies. Greater state investment in their R&D and infrastructure would conflict with the dominant neoliberal regime of market competition among private-sector actors. Meanwhile climate engineering technologies carry a dubious promise, which provides a dual decoy: a public debate over their biophysical risks, and a 'last-resort' alibi for system continuity.

By analogy with earlier pollution technofixes, CCS promises to securely contain GHG emissions by shifting the burden across space and time, thus eliciting wishful responses and widespread rhetorical support. Like climate engineering technologies, 'clean' fossil fuel technology has a socio-political misalignment with the neoliberal regime. Few state bodies have made a significant financial investment in CCS, which is weakly conducive to market-type instruments. 'Given the neoliberal innovation regime's inability to produce implementation via competition and markets, continued commitment to the promise of CCS has been necessary, as a response to the misalignments' (Markusson et al, 2017: 7). Many governments continue their rhetorical commitments to CCS, essential for imagining fossil fuels as eventually low carbon and so justifying their long-term lucrative role (see Figure 2.1).

Nuclear power likewise has a misalignment with neoliberal policies. When the UK government sought to expand nuclear power around 2010, its costs were becoming economically uncompetitive with truly renewable sources. This conflicted with the UK government's ultra-neoliberal framework for energy market competition. Yet the government strongly promoted and subsidized a new nuclear plant at Hinckley Point C. This anomaly

Figure 2.1: Elusive carbon capture

Source: Cathy Wilcox, included with permission of the artist, www.cathywilcox.com.au/

can be explained by an implicit policy aim, namely: providing plutonium for nuclear-propelled submarines to carry nuclear weapons. This policy commitment has had the pretence of an 'independent' nuclear capacity, helping the UK to maintain a strong place in global affairs (Cox et al, 2016). This implicit policy driver explains an exception to the general pattern of technofixes being selectively promoted to fit market-type instruments.

Those patterns illustrate how any technofix has a sociotechnical character, facilitating or favouring a specific form of social order. In the late 20th century, evolutionary economists identified correspondences between each major technological change and its 'techno-economic paradigm'; but this concept was criticized for naturalizing societal change (Freeman, 1991). Later writers used the concept to identify how a technoscientific 'common sense' influenced institutional and social change (Perez, 2009; Drechsler, 2011). By contrast with some techno-economic paradigms, a techno-market framework favours technofixes amenable to market-type relationships. This returns us to an earlier question: how and why policy agendas favour some technofix options over others.

Techno-market fix as a policy framework: its origins and roles

A techno-market policy framework seeks to create new markets whose competitive forces will stimulate eco-efficient technological solutions.

This policy framework arose from merging two antecedents, ecological modernization and neoliberal environmentalism. This section explains them in turn and then their relevance to the EU, the political context for most case studies here.

Ecological modernization

In the 1980s environmentalist protest against harmful production systems generated political demands to reduce their resource burdens and to substitute more environmentally sustainable methods. The latter were often framed as conflicting with economic growth and competitiveness. A novel policy framework sought to reconcile those aims through eco-efficient innovation. This was conceptualized as ecological modernization (ecomodernism), which became a basis for new multi-stakeholder coalitions.

In ecomodernist policy frameworks, states aim to stimulate better self-regulation of industry, thus transferring responsibilities to the market (Mol, 1996: 306). Such agendas have promoted institutional changes to facilitate eco-efficient technofixes. As a critical analyst noted, ecomodernism 'uses the language of business and conceptualizes environmental pollution as a matter of inefficiency, while operating within the boundaries of cost-effectiveness and administrative efficiency'. This is 'basically a modernist and technocratic approach to the environment that suggests that there is a techno–institutionalist fix for the present problems' (Hajer, 1995: 31–32).

Ecomodernism has been sarcastically described as 'the gospel of eco-efficiency'. This dual economic-engineering approach places faith in technoscientific advance which can more efficiently use natural resources, materials and energy. This approach aims to delink economic growth from its material base in resource burdens (Martinez-Alier, 2002: 5–8).

Such eco-efficiency promises underlay many pollution-control technologies. Businesses with polluting processes have sought to reduce costs and avoid financial liability. They generally have devised technofixes for three types of displacement: transformation, relocation and time-delay. In the transformational technofix, a polluting by-product is transformed into a useful, saleable commodity, for example, recyclable plastic. Such techniques are closely linked to the relocational technofix, selling the waste or paying for its transfer elsewhere. Those fixes often have unintended consequences: environmental degradation may be simply shifted a generation or two into the future (Lecain, 2004). For example, exported plastic waste turns out to be difficult to recycle, imposing extra environmental and health burdens. Such displacements exemplify wider means of managing capitalist crisis through temporal and spatial displacements (Harvey, 2018).

Such fixes can be ambiguous, even illusory, though wishfully attractive. These displacements are illustrated by CCS: 'Threaded through the reverie for carbon capture is the fantasy of industrial absolution – that a technology could almost be dreamed into being that could purify the ecological legacy of modernity, even perhaps eliminate its footprint entirely' (Wallace-Wells, 2019: 181). This illustrates wishful narratives which characterize many technofixes and hence their broad appeal. As the state has funded R&D on carbon abatement technologies, 'each technological promise has enabled a continued politics of prevarication and inadequate action, by raising expectations of more effective policy options becoming available in the future, in turn justifying existing limited and gradualist policy choices …' (McLaren and Markusson, 2020). Such promises contribute to high-carbon system continuity.

In an ecomodernist framework, capitalist modernization is seen as benign: societal progress is hindered mainly by market failure to provide appropriate incentives, whereby markets can be readily corrected. From this problem-diagnosis, solutions warrant state measures to adjust the market with appropriate incentives. As critics have argued, this policy framework is limited by 'the preoccupation with efficiency and pollution control over broader concerns about aggregate resource consumption and its environmental impacts' (Buttel, 2000: 64). Through this framework, dominant technology experts often have marginalized environmentalist groups and their critical experts.

Neoliberal environmentalism

As a parallel development in the 1990s, environmental policy was incorporating neoliberal assumptions. As historical background, 18th-century classical liberalism had combined laissez-faire with state coercion to privatize commons, turn land and resources into private property, and transfer its ownership to elites. Going beyond laissez-faire in the late 20th century, neoliberal regimes established financial instruments which deepened market-competitive relations through contracts of various kinds. Such frameworks have promoted individual liberties and property rights as the basis for rationally pursuing one's sovereign self-interest. Fairness is understood as 'the rational agreement of agents to cooperate with one another to further their self-interest' through supposedly voluntary contracts (Okereke, 2007: 43). In practice these arrangements often have depended on state coercion, reinforcing inequities of resources and power between states, as well as among people.

With that wider framework, some market-type instruments have been promoted as necessary for incorporating environmental concerns into economic practices. This framework has been conceptualized as neoliberal or market environmentalism (Bernstein, 2001; also Bailey, 2007). Environmental policy was turned into opportunities for new markets as necessary instruments.

By the 1990s, ecomodernist policy frameworks were merging with neoliberal environmentalism, most notably in the 1997 Kyoto Protocol under the UNFCCC. Here the state created and subsidized carbon credits as market-type instruments. Carbon trading was meant to incentivize eco-efficient innovation which will strengthen environmental protection and economic competitiveness. Despite the official aims, tradeable credits have provided means for companies to appropriate natural resources, while disguising this as a free contractual exchange. As a techno-institutionalist fix, ecomodernism has favoured stakeholder groups who are deemed to hold expert knowledge, especially market actors, while marginalizing low-income and civil society groups from the global South. The system 'is largely reduced to cheerleading for private and voluntary national action on climate change' (Ciplet and Roberts, 2017: 154).

The convergent framework is a techno-market fix, potentially extending or simulating competitive market relations. As David Harvey argued, the neoliberal drive to extend markets has been linked with the technological fix, which 'relies on the coercive powers of competition'. This 'becomes so deeply embedded in entrepreneurial common sense, however, that it becomes a fetish belief that there is a technological fix for each and every problem' (Harvey, 2005: 68).

Despite the rhetoric of free-market competition, neoliberal agendas sometimes have generated novel regulatory forms through new organizations, strategies and evaluation criteria (Busch, 2010: 334). Such 'neo-regulation' has helped to adjust market competition in ways which can legitimize new markets, sometimes in response to dissent. The EU has expert-regulatory procedures potentially channelling and absorbing dissent (as shown in Chapters 3 and 4).

Societal futures depoliticized

In EU policy a techno-market framework emerged from several historical shifts. The European integration project originated as an expert-based agenda to eliminate trade barriers across member states. Even before the 1970s, trans-European expert networks had framed many issues as 'technical', seeking common standards which could overcome cross-border trade barriers and political conflicts alike. These efforts provided an expert basis for Jean Monnet to propose 'a transfer of power to common institutions, majority rule and a common approach to finding a solution to problems' (Monnet, 1974). Success depended on a Europe-wide 'functional integration' through expert committees, later called the 'Monnet method'.

That administrative method was later extended to economic and political integration by portraying all trade barriers as technical issues. Moreover, the EU embraced technoscientific innovation as a collective identity and political mission, yet supposedly standing above politics (Kaiser and Schot, 2014). This agenda aspired to global leadership in ecomodernist techno-innovation.

Since at least the 1990s, the EU's ecomodernist framework has emphasized eco-efficiency measures to reconcile economic growth with environmental sustainability. As an ongoing tension, EU has made symbolic declarations on 'sustainable development', whose meanings have come from ecomodernist strategies. These have complemented the European integration project to create a barrier-free internal market.

Sustainable development has been understood mainly as market-based eco-efficiency measures decoupling economic growth from environmental harm. Nature has been framed as a 'standing reserve' of exploitable resources. Ecomodernism has served 'the construction of a neoliberal free-market economy in support of industrial competitiveness' (Baker, 2007: 302–303).

Illustrating this ecomodernist framework, from the 1980s onwards the European Commission's R&D programme has sought eco-efficient biotechnological innovation for using natural resources more productively (Gottweis, 1998: 232–235; see also Chapter 3). After mass protest deterred a GM products markets, in 2005 agribiotech was relaunched as the KBBE, recasting agri-production as biomass which could be genetically recomposed into end-products functionally substituting for fossil fuels (see Chapter 4). In 2010 the EU was rebranded as an 'Innovation Union', prioritizing more efficient production methods for increasing economic competitiveness and achieving a more 'eco-efficient economy' (Lund Declaration, 2009; CEC, 2010a, 2010b).

On that basis the EU has elaborated a future narrative of societal progress. According to critics: 'Science and technology in this imaginary are staged unambiguously as the solution to a range of social ills, including the problematic identity of Europe itself. To the extent that S&T are recognised to generate problems, these are cast solely in the form of mistaken technological choices' (Felt et al, 2007: 80). Hence political debate is reduced to the search for more effective technological solutions and market incentives for their development (van den Hove et al., 2012).

EU policy frameworks have anticipated and promoted technoscientific development as central to societal progress, thus depoliticizing policy choices and responsibility for them. Europe has been cast as a potential loser or winner in a global race to commercialize technoscience, as a crucial means to address societal problems. Nevertheless recurrent public dissent has often reopened 'technical' issues as political ones, pressed state bodies to defend their versions of the public good and counterposed rival futures, as explained next.

Rival societal futures

In conflicts over environmental issues, rival agendas promote different visions of a feasible, desirable future. Their rivalry generates diverse meanings of key terms such as sustainable, green, low-carbon, circular economy, and so on. Let us paraphrase a famous literary scene by Lewis Carroll (1871):

Humpty Dumpty:	When I use a word, it means just what I choose it to mean – neither more nor less.
Alice:	The question is whether you can make words mean so many different things.
Humpty Dumpty:	The question is, which [who] is to be master – that's all.

What enables the mastery of meanings? By whom? How do rival futures contend for influence? These have been theorized as imaginaries, which are explained here.

Sociotechnical imaginaries

Future societal visions entail economic and sociotechnical imaginaries. By mobilizing symbolic and political support, a future vision can play a self-fulfilling performative role. These concepts help explain how some potential futures 'are selected, get embodied in individual agents or are routinized in organizational operations' (Jessop, 2010: 339).

Economic imaginaries have been conceptualized by cultural political economy. An 'imagined economic space' may become grounded in an 'imagined community of economic interest' (Jessop, 2005: 162). In order to assemble effective coalitions, actors 'articulate strategies, projects and visions oriented to these imagined economies' (Jessop, 2010: 345).

In their performative role, imaginaries serve to mobilize economic resources and political support, thus creating the conditions to achieve specific futures. Economic imaginaries often frame territorial jurisdictions as competitive units in a global economic rivalry. As the dominant economic imaginary since the 1990s, the nation (or 'Europe') becomes a single competitive space facing a common external threat and market opportunity (Rosamond, 2002: 169).

Science and Technology Studies (STS) has developed a complementary concept: sociotechnical imaginaries, which articulate 'the relationship of science and technology to political institutions'. They are 'collectively imagined forms of social life and social order reflected in the design and fulfilment of nation-specific scientific and/or technological projects' (Jasanoff and Kim, 2009: 120). The adjective 'sociotechnical' identifies links various social aspects such as technological design and state support measures for a specific societal order.

Such imaginaries inform innovation policies:

> Such policies balance distinctive national visions of desirable futures driven by science and technology against fears of either not realizing those futures or causing unintended harm in the pursuit of technological advances. S&T policies thus provide unique sites for exploring the role of political culture and practices in stabilizing particular imaginaries. (Jasanoff and Kim, 2009: 121)

An imaginary may serve to mobilize desires, organizations and resources, thus shaping societal futures:

> By turning to sociotechnical imaginaries, we can engage directly with the ways in which people's hopes and desires for the future – their sense of self and their passion for how things ought to be – get bound up with the hard stuff of past achievements, whether the material infrastructures of roads, power plants, and the security state, or the normative infrastructures of constitutional principles, juridical practices, and public reason. (Jasanoff, 2015: 22)

The concept helps to understand 'why different moral valences attach to new scientific ideas and technological inventions'; likewise how actors 'construct stories of progress in their programmatic statements, and how they blend into these their expectations of science and technology' (Jasanoff, 2015: 25). The dominant imaginary can stimulate alternatives, which may be conflictual or complementary: 'Multiple imaginaries can coexist within a society in tension or in a productive dialectical relationship' (Jasanoff and Kim, 2015: 4).

As a prevalent form of sociotechnical imaginary: 'The techno-market imaginary ... assumes that the environment is somewhat vulnerable, but that the climate issue is manageable through appropriate economic incentives and technological innovation, without fundamentally compromising lifestyles or economic growth. This imaginary's positioning highlights its hegemonic appeal, by claiming to reconcile economic and environmental concerns' (Levy and Spicer, 2013: 666). The **techno-market imaginary** 'allocates a primary role to the private sector in addressing climate change, lending this imaginary a broad appeal across multiple constituencies'. This framework promotes financial instruments, especially carbon trading, to incentivize techno-innovation (Levy and Spicer, 2013: 664, 669).

The concept 'sociotechnical imaginaries' built on earlier perspectives about any innovation co-produces complementary forms of nature, technoscientific knowledge and social order. The ways in which we know and represent the world are inseparable from the ways in which we live in it (Jasanoff, 2004: 2–3). Through **co-production**, a sociotechnical arrangement constructs social order in particular ways, while attributing that order to natural or technical characteristics (Jasanoff, 2004: 21). By distinguishing 'technical' characteristics from 'social' ones, an agenda can normalize a specific social order as if it had an apolitical value-free basis. Yet such efforts often provoke dissent, highlighting the value-laden social basis of technical characteristics, which readily become contentious. This analytical perspective helps to analyse how a specific social order is promoted, stabilized or destabilized.

Social agency as a transformative mobilization

The ecomodernist agenda has provoked protest, generating or highlighting alternatives. Yet these often remain as technical-administrative blueprints, or as appeals to hypothetical planners having the will, capacity and resources to implement them. In particular, renewable energy proposals have appealed to imaginary planners (Lohmann and Hildyard, 2013). Even when implemented, such initiatives often remain as supplements to dominant high-carbon systems rather than as replacements for fossil fuels. For alternatives to bring system change, an effective social agency depends on a transformative mobilization integrating five main elements: mobilized counter-publics, frame alignments, an eco-localization imaginary, grassroots innovation and solidaristic **commoning**. Those elements are briefly explained here.

Alternative agendas have come from multi-stakeholder citizen–expert alliances, sometimes involving small-scale producers. Together they have contested official knowledge-claims about the benefits or disadvantages of the dominant innovation agenda. Such opposition has drawn on knowledge from socially excluded groups (for example, service users, patients, low-income groups, and so on), facilitated by NGOs and social movements. Such 'mobilised counter-publics' often have generated public controversy over dominant agendas, prevented public consent and counterposed transformative futures (Hess, 2007, 2016).

Criticizing dominant policy assumptions, such protest has highlighted the anti-democratic basis of technicized decision-making. Counter-publics sometimes identify 'undone science'; they demand or generate resources for new knowledge which could serve a broad public benefit rather than private interests. Some mobilize resources to fill the knowledge gap (Frickel et al, 2010).

Counter-publics often emerge from social movements, whose participants bring diverse framings of a societal problem, for example, environmental or health threats, socio-economic inequity, resource degradation, and so on. Effective action depends on integrating them for and through common action. As a feature of social movements, 'frame-bridging' aligns 'two or more ideologically congruent but structurally unconnected frames regarding a particular issue or problem' (Snow et al, 1986: 467; see also Snow and Benford, 2000: 624). Such alignments strengthen the basis for jointly advocating an alternative future.

As a general example of frame alignment, the sufficiency principle or sufficiency economy has been counterposed to economic growth. According to a labour movement network, for example, 'trade unions must develop transformational strategies that are anchored in a paradigm of sharing, solidarity, and sufficiency' (TUED, 2018: 43; see also Chapter 6). The concept has been elaborated elsewhere as follows: 'Economies should seek to universalise a material standard of living that is sufficient for a good life but which is ecologically sustainable into the deep future', encompassing many possible forms (Alexander, 2015: xiv, xix).

Alternative economic models can be understood more broadly as eco-localization. This may be simply an incidental effect of investors cheapening supply chains on a more local basis. By contrast, when stimulated by climate and social justice issues, eco-localization can express a social movement; this builds more enjoyable lives by creating lower energy forms of livelihoods and localizing production-consumption circuits, even if resource-poor by conventional criteria. In such ways, 'advocates of intentional localisation are developing radical new conceptions of livelihood and economy that directly cut against the logic of growth-based capitalist economic strategies and elite conceptualisations' (North, 2010: 586).

When it becomes an explicit transformational agenda, eco-localization promotes local economies that are socially just, diverse and resilient. To alleviate socio-economic inequalities, they may demand resource transfers from affluent to low-income areas (North, 2010: 588). Such agendas facilitate participatory experimentation towards more effective practices, alongside interdependent learning processes among localities (North, 2010: 587; also Levidow and Papaioannou, 2016; Feola and Jaworska, 2019). An urban context encompasses more dense networks and resources, generating diverse political actions; they can jointly sustain grassroots interventions that can achieve a sustainability transition (North and Longhurst, 2013).

Such alternatives conflict with mainstream capital-intensive technologies, products and processes. The latter are generally top-down, expensive capital-intensive innovations which exclude lower-income people, their interests and aspirations for better lives (Cozzens and Kaplinsky, 2009). Alternatives reframe innovation in socially inclusive ways. After the 'inclusive innovation' concept emerged from the global South (Gupta, 1996), it was soon seen as relevant globally, especially after the 2008 financial crash. The most developed countries had competitiveness policies which worsened their domestic poverty problems. As a remedy, inclusive innovation should mean greater rights, voice, capabilities and incentives for excluded people. On this basis, they can become active participants in development and innovation processes (Johnson and Andersen, 2012: 7–8).

As an umbrella concept, inclusive innovation encompasses diverse forms. A greater emancipatory potential comes from grassroots innovation, that is, driven by innovators' capacities, knowledge exchange, uses and benefits (Levidow and Papaioannou, 2018). These processes respond to social injustices, socio-economic inequalities and environmental problems. Innovators become closely linked with users or become users. Innovations are designed to be socially shared, replicated and spread at low cost (for example, Smith et al, 2014).

Grassroots innovation expands a knowledge base mobilizing supportive resources among wider publics; these involve local actors with different forms of knowledge. Widespread examples include: open-source software being progressively improved; agroecological methods being enhanced through producers' knowledge-exchange; community-owned renewable energy; and FabLabs creating flexible tools. The latter case has generated practical questions such as 'What tools for what social purposes?' Hence grassroots innovation readily becomes 'entangled in the politics of technology' (Smith et al, 2015: 10).

For a grassroots innovation to multiply and spread, participants need to integrate diverse aims and capacities. This integration depends on solidaristic reciprocity and learning processes (Feola and Butt, 2017). However, grassroots innovation encounters barriers or diversions in the wider economic system. More inclusive grassroots processes achieve greater participation but may lack the power to reshape innovation agendas. They can either adapt to the mainstream, or else mobilize resistance to the mainstream through a transformative agenda (Fressoli et al, 2014: 12). Initiatives may face tensions between those trajectories.

For a transformative role, grassroots innovation depends on a cooperative, commons-based process: 'Recognising and resourcing the value of commoning in technology involves struggle – with funders, with evaluators, with investors, and against predatory encroachments. That is because ultimately it involves a deeper redistribution of capabilities: the capability to challenge power by redefining and redistributing political and economic resources to these initiatives' (Smith et al, 2019).

Such a process would need to extend commons, while protecting them against enclosure. That phrase originates from England's 18th-century land enclosures. Since the late 20th century, an analogous neoliberal process has been enclosing a wider range of commons. These consist of a shared interest or value that is produced through cooperative relations within a community. Commons may provide access to food, water, forests and clean air on a basis other than market relations (Di Chiro, 1998). For populations marginalized by development, a community sustains and manages commons to provide crucial everyday means of social reproduction, for example, food, education, health and/or welfare (Caffentzis, 2016: 100).

Across its diverse forms, the term 'commoning' describes a cooperative process – 'thinking, learning and acting as a commoner' – beyond simply managing resources together (Bollier and Helfrich, 2015: 2–3; see also Wall, 2014). This process protects, constructs and/or enhances resources as a commons. This arrangement is managed by a solidaristic community which protects resources from degradation and market competition.

> Commons are necessarily created and sustained by communities, i.e. by social networks of mutual aid, solidarity, and practices of human exchange that are not reduced to the market form. The place of these networks does not need to be tied to locality, but communities can operate both in local and through trans-local places. (de Angelis, 2003: 1; also 2017)

In sum: To decarbonize economies in socially just ways needs transformative mobilizations combining the five aspects discussed here: mobilized counter-publics, frame alignments, an eco-localization imaginary, grassroots innovation and solidaristic commoning. In pervasive conflicts between techno-market fixes versus alternatives, each agenda promotes its own imaginary to gain support and shape the future. Each one frames natural resources, knowledge and social order in a distinctive form. Together those concepts will help to illuminate the case studies here, as a basis to draw general insights from them.

Participatory Action Research for social agency

Alongside the concepts already discussed, a crucial method has been Participatory Action Research (PAR). Put simply, this means research with people rather than on them; its 'extended epistemology' encompasses experiential and practical knowing (Heron and Reason, 2006, 2008). PAR brings together researchers with practitioners, initially to identify practical problems and analytical questions that warrant joint research.

PAR originated in Latin American struggles against the hegemonic power with its everyday neocolonial oppression and exploitation. There the research method has had various forms and names, especially Investigación Acción Participativa as an umbrella concept (for English-language translations, see Fals Borda, 2001; Coghlan and Brydon, 2014). The methods were adapted elsewhere by researchers bringing social justice perspectives to political-economic conflicts.

Some participatory forms (including PAR) were eventually reduced to yet another data-collection technique, or they were assimilated and mainstreamed into neoliberal agendas. This shift prompted calls to repoliticize PAR for its original counter-hegemonic, emancipatory aims (Jordan and Kapoor, 2016). This caveat remains relevant for any research partnership.

Through PAR, participants should become empowered to play the role of change agents. Creating conditions for such partnership helps 'generate a sense of connection that better ensures benefit of the collective good', initially among participants (Bradbury, 2010: 98, 106). The basic concept has been elaborated in various ways appropriate to actors' aims in specific contexts.

Research collaboration has been sometimes structured as a series of steps or phases. These facilitate systematic group learning and knowledge-sharing; they adjust actions for alignment with agreed objectives, while empowering the actors to learn and adapt their strategies (Lewin, 1946). In one version, a reconnaissance step initially explores the situation, then an intervention seeks to improve the situation, and finally a reflection evaluates practical outcomes vis-à-vis the initial aims and research questions. This provides a basis to plan the next cycle (Melrose, 2001). In those ways, '[a]ction researchers plan for cycles of action and reflection and thereby must be *reflexive* about how change efforts are unfolding, and the impact that our presence (the intervention) is having. ... Researchers need to discuss and shape our research question and design with the practitioners' (Bradbury, 2010: 98).

Social movements have a fluidity and internal diversity which may be awkward for a formal research collaboration, much less for a step-wise process. Some researcher-activists have directly participated in social movements as a PAR method. When engaging with the Climate Justice movement, some called their participant role an 'accompaniment', meaning an internal collaboration based on trust, with a common understanding of the problem that warrants research. They investigated various strategies to contest the UN Climate Convention, triggering diverse reactions from climate activists. Some mainstream groups eventually echoed criticisms from the Climate Justice movement and distanced themselves from market-based solutions of the Climate Convention (Reitan and Gibson, 2012). Thus researcher-activists brought insider insights about significant political shifts, as mentioned earlier (for example, Hadden, 2015; della Porta and Parks, 2017).

Regardless of the specific form, PAR has two levels: researchers intervening in stakeholders' practices, at the same as they intervene in a wider context. Through this collaborative relationship, participant groups can gain a better collective self-understanding of their problems and opportunities, as a basis for more effectively addressing them. This process can strengthen social agency for transformative aims.

As the general policy context of environmental technofixes, state bodies have framed risk or sustainability issues as direct biophysical effects of a product or technology. This frame has potentially channelled dissent into specialist issues, thus obscuring systemic drivers of harm. Regulatory procedures have evaluated potential effects through implicit assumptions as regards what potential effects may be relevant, acceptable or better than some standard. Counter-publics have often questioned those criteria, thus extending public controversy to regulatory expertise.

Moreover, they have turned such disputes into opportunities for highlighting political-economic interests and institutional commitments driving the fix. Whether formal or informal, PAR processes have clarified such strategies of counter-publics, as analysed in each case-study chapter. The Conclusion will recapitulate these strategies for their broader relevance.

3

EU Agribiotech Fix:
Stimulating Blockages and
Agroecological Alternatives

Introduction

Climate change was becoming a more salient issue in the run-up to the 2014
UN Climate Summit and the 2015 Paris COP21. Using the opportunity,
an industry–state alliance promoted 'climate-smart agriculture', emphasizing
capital-intensive inputs such as GM herbicide-tolerant crops for no-till
agriculture. Critics denounced this proposal as a 'false solution', even as
'corporate-smart greenwash'. They counterposed agroecology as a truly
climate-resilient agriculture. This dispute extended a decades-long conflict
over technofixes for endemic harms from the agri–industrial system.

By the 1960s, intensive pesticide usage was inflicting serious environmental
harms, especially in North America. Rachel Carson's *Silent Spring* gained
prominence for popularizing the dangers and scientific evidence. Her book
denounced pesticides as 'biocides' that kill potentially all insects or weeds,
thus destroying the basis of life. More fundamentally, her book diagnosed
a political-economic driver of the problem, namely: the US's agricultural
over-production, a policy to reduce the acreage in production, a financial
incentive to maximize yield on the smaller acreage and greater agrichemical
usage, thus degrading biodiversity (Carson, 1962: 19).

Widespread protest demanded restrictions or bans on many pesticides,
while agrichemical companies denied the dangers. Some eventually devised
genomics techniques for a new agribiotech industry, later called the Life
Sciences. This promised genetic solutions that would better protect crops
and reduce farmers' dependence on agrichemicals.

What systemic problem was at stake? Plant breeding had traditionally
provided some pest resistance through selection from a crop's diverse
characteristics. But in the 1970s–1980s the process came 'under the pesticide
umbrella', relying on pesticides as a basis to prioritize yield, thus weakening

intrinsic pest resistance. The greater vulnerability was then portrayed as natural defect which must be corrected by GM techniques. Thus the GM agribiotech fix arose from problems of the previous fix.

The European Community (EC) embraced such potential solutions from the 1980s onwards. When the first agribiotech (GM) products were being marketed in the late 1990s, however, opponents launched a Europe-wide protest campaign. This stigmatized GM products by association with the 'mad cow' pandemic, factory farming and its systemic hazards. Diverse groups converged into an effective opposition, eventually blocking a European market for GM products. Critics broadened their target: agribiotech symbolized an unsustainable agri-industrial system. Moreover, agribiotech opponents used the European controversy as an extra opportunity for various 'quality' alternatives.

The European GM controversy still has global reverberations for agri-food innovation and new technologies in general. Opponents' success encouraged self-confidence and sharper strategies for resisting other undesirable innovations. By contrast, some techno-optimists sought means 'to avoid another GM debacle'; they saw the European blockages as exploiting irrational fears, politicizing science and impeding technological progress.

Did agribiotech critics truly politicize science? This elite narrative inverts reality. On the contrary, the agribiotech lobby scientized public policy by portraying the issues as merely technical; this marginalized less powerful actors and their alternative agendas, at least initially (Kinchy, 2012). The lobby sought to depoliticize its neoliberal agenda and thus pre-empt or disguise the societal choices at stake.

Public controversy arose partly from the agribiotech policy framework and innovation trajectory, which appropriated genetic commons as intellectual property, while also degrading biodiversity. Activist networks highlighted such threats in order to intensify and globalize the controversy. They used various political and market opportunities to block GM products (Schurman and Munro, 2010).

In the EU context, the controversy and blockages opened up greater opportunity for 'quality' alternatives to the agri-industrial system. To illuminate such dynamics, this chapter analyses systemic sources of the recurrent EU-wide conflict from the late 1990s onwards. It draws on informal kinds of Participatory Action Research (PAR), which brought together academics with civil society groups highlighting political-economic drivers of the agribiotech agenda.

Extending that inquiry, this chapter discusses the following questions:

- How did the EU promote agribiotech through institutional change and a future societal vision?
- How did protest help to block GM products?

- How did opponents create and use a greater opportunity for 'quality' alternatives, eventually agroecology?
- How do those divergent agendas relate to system change versus continuity?

To answer those questions, this chapter analyses institutional commitments and techno-optimistic assumptions underlying the EU's agribiotech fix. In particular, the policy framework promoted several market-type drivers as conditions that were supposedly necessary to bring societal benefits. The analysis elaborates the concept 'techno-market fix' as a policy framework that combines ecological modernization with neoliberal environmentalism. Likewise rival sociotechnical imaginaries underlay the agribiotech fix versus agroecological alternatives (see again Chapter 3).

The chapter has the following structure: the first section on how the EU's policy framework elaborated agribiotech through a techno-market fix; the second section on how such changes provoked the 1990s controversy, bringing together diverse opponents; the third section on how more stringent regulation was used to block GM products and promote quality alternatives; the fourth section on how opponents promoted an agroecology agenda for transforming the agri-food system; the fifth section on how global disputes arose over agribiotech for 'climate-resilient agriculture'; and, finally, the sixth sections returns to the questions posed in the list. Table 3.1 summarizes rival sociotechnical imaginaries underlying the conflict.

Table 3.1: Rival sociotechnical imaginaries: Life Sciences techno-market fix versus eco-localization

Sociotechnical imaginary*	Life Sciences techno-market fix	Eco-localization via agroecological systems
Problem-diagnosis: threats to be overcome	Inefficient production methods (from deficient crops) disadvantage European agro-industry, which thereby falls behind in the global market competition for technoscientific advance.	Agri-industrial monoculture systems make farmers dependent on external inputs, undermine their knowledge, distance them from consumers and degrade food quality. These destructive systems have various remedies.
Techno-scientific knowledge and access	'Clean' genetic changes provide more eco-efficient crops enhancing economic competitiveness. Scientific research designs novel inputs for transfer to farmers.	Agroecological innovation through farmers' knowledge exchange and commons on natural resources. Scientific research tries to explain why some agroecological practices are more effective.

(continued)

37

Table 3.1: Rival sociotechnical imaginaries: Life Sciences techno-market fix versus eco-localization (continued)

Sociotechnical imaginary*	Life Sciences techno-market fix	Eco-localization via agroecological systems
Agri-food innovation (Vanloqueren and Baret, 2009)	Life Sciences engineering attempts to substitute biological inputs for agrichemicals, and to diversify outputs of plant-cell factories.	Agroecological methods aim to enhance agroecosystem biodiversity as a means to improve crop protection, productivity, nutritional quality and resource conservation.
Nature and knowledge. Agricultural environments	Deficient crops needing genetically precise improvements drawing on molecular-level databases. Homogenized environments for regulatory harmonization.	Agroecosystems optimizing ecological interactions and farmers' knowledge of them, within and beyond agricultural fields. Value biodiversity in evaluating potential effects and wider benefits.
Intellectual property	Genetically precise changes are biotechnological inventions warranting broader patent rights to protect them from biopiracy, that is, unauthorized use.	'Patents on Life' are biopiracy: they privatize the genetic commons, make seeds more homogeneous and marginalize farmers' knowledge of natural resources.
'Adverse effects' to be prevented	Only harm worse than normal hazards of intensive monoculture.	Any harm worse than chemical-free cultivation methods, for example, herbicide effects and 'genetic contamination' (which served as a bridge frame).
Coexistence of GM and non-GM crops	Regulations should protect non-GM crops from any (evidenced) biophysical risks of GM material.	Regulations should protect non-GM crops from GM contamination and reputational threats from nearby GM crops.
Product quality (Allaire and Woolf, 2004)	Genetically precise compositional changes conferring extra market value, for example, pesticidal or organoleptic characteristics.	Integral product identity via holistic methods and quality characteristics recognizable by consumers, as a basis for their support.
Social order	Market competition for eco-efficient innovation serving more environmentally sustainable agri-production methods.	Solidaristic relations among producers and with consumer-citizens. Cooperative relationships exchanging knowledge about biodiverse agri-production methods.
Climate-resilient agriculture	'Climate-smart agriculture' would use various inputs to strengthen resilience against climate stresses. Herbicide-tolerant crops avoid tilling the soil and so help conserve it.	Agroecological climate-resilient methods avoid external inputs incurring GHG emissions, store carbon in organic matter, conserve biodiverse soil, complement no-till and so 'cool the planet'.

* Each sociotechnical imaginary co-produces the distinctive forms in the rows.

Note on terminology: Across Europe's languages, the technology or its products are variously called genetic modification (GM), OGM (French), *Genteknik* (German), *plantas transgénicas* (Spanish and Portuguese), and so on. The industrial sector has been called agricultural biotechnology or agribiotech for short. For simplicity, the terms GM and agribiotech will be used here.

EU's agribiotech policy as a techno-market fix

In the 1980s the European Commission was promoting agribiotechnology as essential to enhance eco-efficient agri-production methods, protect the environment and make European agriculture more globally competitive. It advocated a 'clean technology' base, which would facilitate a positive relation between the environment and economic growth. Its policy counselled European adaptation to inexorable competitive pressures: 'The pressure of the market-place is spreading and growing, obliging businesses to exploit every opportunity available to increase productivity and efficiency' (CEC, 1993a: 92–93).

This imperative justified support for innovations such as biotechnology: 'The European Union must harness these new technologies at the core of the knowledge-based economy' (CEC, 1993a: 7). Likewise the 5th Environmental Action Programme celebrated new technologies which could provide eco-efficiency gains towards environmental objectives (CEC, 1993b: 28). EU policy echoed claims from agribiotech-agrochemical companies that GM crops simultaneously achieve economic and environmental objectives, especially by reducing input costs and waste (Monsanto, 1997: 16). Early critics warned that these new agri-industrial inputs would undermine traditional agricultures and natural resources; lower prices could intensify farmers' competition for higher yields.

The agribiotech industry anticipated and emphasized greater market-competitive pressures on European agriculture, as both an imperative and opportunity for their solutions. Under reforms of the Common Agricultural Policy (CAP), EC agricultural subsidies would be reduced and would lose their former link with production. In the view of many company managers, market liberalization and subsidy reduction would continue, thus offering greater opportunity to sell inputs to farmers, as well as to finance future development of GM crops (Chataway et al, 2004: 1053). Thus market-driven eco-efficiency solutions were promoted in the name of helping farmers to compete in global markets.

Farmers' traditional seed-saving had maintained a knowledge-skills commons, thus limiting private appropriation and capital accumulation by input suppliers. To overcome this barrier, intellectual property rights (IPR) legislation aimed to enclose the commons of biodiverse seeds and farmers' knowledge (Mascarenhas and Busch, 2006; Parfitt, 2013). Along those lines,

Figure 3.1: Criminalize Biopiracy

Source: ASEED

in 1988 agribiotech supporters proposed an EC Directive extending patent rights to 'biotechnological inventions', thus turning genetic-level discoveries into private property.

This proposal provoked a public controversy over 'biopiracy', which had two contrary meanings. For the patent-holder, it meant the unauthorized use of GM seeds; for critics, it meant illicit 'Patents on Life', that is, patent rights on mere discoveries of common resources (see Figure 3.1). The 'biopiracy' issue raised doubts among those in the political Left and tradeunions which were otherwise inclined to support technological innovation as societal progress. Given such opposition, the new Directive became law a decade later (EC, 1998).

This broadened the scope of discoveries or techniques which could be privatized. As public controversy continued afterwards, several EU member states failed or refused to incorporate the Directive into national law. Nevertheless the Directive strengthened incentives for company patent holders to use GM techniques in novel seeds.

Public-sector R&D policies did likewise. In many EU member states, agricultural research institutes were allocated less state funds than before and were expected to substitute such income from the private sector or from royalties on patents. The EU's R&D funding priorities complemented that shift towards marketizing hitherto 'public-sector' research, thus blurring the public/private distinction (Levidow et al, 2002). By 1990 EC funds for biotech research became conditional upon industry partners committing resources to a project.

Research was given a clear economic role, with 'more careful attention to the long-term needs of industry', according to managers of the DG Research Biotechnology Division (Magnien and de Nettancourt, 1993: 51). In their view, 'The most vital resource for the competitiveness of the biotechnology industry is the capacity to uncover the mechanisms of biological processes and figure out the blueprint of living matter' (Magnien and de Nettancourt, 1993: 53). As an information machine, Nature had deficiencies which must be identified and corrected in order to strengthen European competitive advantage.

Along those lines, biotechnology was given prominence in the Commission's research framework programmes. The agenda emphasized 'technologies needed to design and develop processes and produce "clean", high-quality products'. GM techniques were promoted in the name of 'pre-competitive' research, whose results could later be developed into marketable products. Framework Programme 5 (1998–2002) included a large programme on 'Life Sciences and Biotechnology', which echoed industry's agenda for synergies between pharmaceutical and agri-food research, sometimes manifest in company mergers (Tait et al, 2002).

Those priorities strengthened drivers to privatize knowledge and resources, while marginalizing farmers' knowledge of natural resources. 'Life Sciences' framed the crucial knowledge as molecular-level databases, necessary for technoscientific innovations that would help to green agriculture and bring novel pharmaceuticals, especially through a convergence of those sectors. As a sociotechnical imaginary, Life Sciences served to justify various neoliberal policies to facilitate innovation for societal progress.

Public controversy over unsustainable agri–industrial systems

As EU policy promoted eco-efficiency benefits for international competitiveness, this political imperative was internalized in the regulatory framework, especially its narrow criteria for harm. As the official framings were contested, however, they were superseded by sustainability and development issues in a broad public controversy. Such disputes both drew on and stimulated an agrarian-based rural development model, enhancing farmers' knowledge of natural resources as a basis for 'quality' products; this rejected the imperatives of globalized agri-food markets. Counter-publics gained this political outcome at several stages, as shown here.

Regulatory criteria disputed

EU member states had a duty to ensure that genetically modified organisms (GMOs) do not cause 'adverse effects', according to the relevant Directive (EEC, 1990). It left open the definition of 'adverse effects' – what would count as harm. Advocates saw this wording as precautionary, that is, anticipating uncertain or unknown risks. For GM products being initially evaluated, however, EU regulatory procedures conveniently applied a narrow definition of 'adverse effects' as a basis to identify no risk.

As critics had warned for a long time, the familiar 'pesticide treadmill' could be supplemented by a genetic treadmill; in particular, insecticidal or herbicide-resistant crops could generate resistant pests. In promoting such crops, regulatory authorities characterized pest resistance as a familiar 'agronomic problem' which also complicates chemical pesticides and therefore would not count as an 'adverse effect'. This judgement initially prevailed in EU-wide decisions on GM crops. Some member states sought more evidence or control measures regarding such treadmill-resistance effects, but their requests were marginalized in the EU-wide procedure (Levidow et al, 1996).

When the European Commission approved some GM crops for commercial cultivation in 1996–1997, safety claims accepted the normal hazards of intensive monoculture such as pest resistance to pesticides. Amidst national differences over defining 'adverse effects', these judgements were levelled down through the EU regulatory procedure. Risk-assessment criteria homogenized national differences and environments, as a convenient basis to identify no risks from GM crops (Levidow et al, 2000).

The European 'mad cow' crisis soon became a reference point for agribiotech. Before that crisis, market pressures had led to a 'more efficient' technological change, inadvertently facilitating the spread of a deadly agent in cows (Jasanoff, 1997). This undermined the credibility of the EU's regulatory oversight for food products, while also aggravating public suspicion towards intensive agricultural methods. When the European Commission approved a GM insecticidal maize in January 1997, its approval was denounced by a wide range of organizations including the European Parliament. The mass media echoed attacks on the Commission for 'recidivism', that is, for repeating its previous crime over safety assurances for British beef. In this way, protest turned agribiotech eco-efficiency claims into an ominous threat.

Extending such analogies, activists warned against systemic threats from GM crops: they would impose 'uncontrollable risks', would spread 'genetic pollution', would extend 'unsustainable' intensive agricultural methods which had already generated agri-food hazards, and would extend corporate power over the agri-food chain (Levidow, 2000). Activists generated a broad opposition among civil society groups.

In those ways, critics challenged the agribiotech promise that more resource-efficient agri-inputs would remedy environmental problems. They also counterposed different agriculture models, for example, organic farming and Integrated Pest Management. Protest was driven mainly by activists from environmentalist and farmer groups. In some countries, for example, France and Italy, small-scale farmers generated mass opposition to GM crops. In response to public controversy, some governments devised a more cautious regulatory approach, going beyond 'risk' to 'sustainability' issues and alternative development models. Four national examples briefly illustrate those popular pressures and policy responses.

In the late 1990s the French biotechnology debate expanded from 'risk' to sustainability issues. Some industrial-type farmers had sought access to GM crops as a means to enhance their economic competitiveness. By contrast, the Confédération Paysanne attacked such products as a multiple threat – to their economic independence, to high-quality French products, to consumer choice and even to democracy. They counterposed their own *paysan savoir-faire*, as a basis for a different societal future (Heller, 2002). Peasant activists destroyed field tests and Syngenta's stocks of GM seeds, leading to a criminal prosecution. The court case was turned into a theatrical trial of the agribiotech industry, thus increasing public support for the opposition campaign.

From a politicized peasant identity, French critics advocated de-intensification measures, based on 'remunerative agricultural prices and sustainable family farming, with multiple benefits for society' (CPE, 2001). Supporting agribiotech in principle, the French government initially led EU-wide approval of a GM herbicide-tolerant oilseed rape (canola), but soon reversed its stance and blocked approval. This regulatory blockage responded to expert concerns about a genetic-pesticide treadmill spreading herbicide tolerance; the decision also accommodated public anxieties and peasant opposition.

In the UK, mass protest regularly destroyed crops in field tests of GM crops, on several grounds: that GM pollen flow threatened nearby organic crops, that 'Gene Dictators' (such as Monsanto) were seeking to control the food supply, and that the UK government's deference to such companies undermined democratic accountability. Public suspicions were echoed widely among civil society groups of many kinds. The Consumers' Association attacked the agri-food industry for its 'unshakeable belief in whizz-bang techniques to conjure up the impossible – food that is safe and nutritious but also cheap enough to beat the global competition' (McKechnie, 1999). This sarcastic comment played on analogies with the 'mad cow' pandemic, which had resulted from a regulatory change lowering the cost of animal feed.

Italian agribiotech opponents sought to protect the agri-food chain as a cultural environment for artisanal methods and local specialty products known as *prodotti tipici*. In the 1990s the Italian Parliament had allocated subsidies to promote such products and foresaw these being displaced or discredited by GM crops. According to a Parliamentary report, the government must 'prevent Italian agriculture from becoming dependent on multinational companies due to the introduction of genetically manipulated seeds'.

The Parliament further declared: When local administrations apply EU legislation on sustainable agriculture, they should link these criteria with a requirement to use only non-GM materials. Such arguments were being adopted from Coltivatori Diretti, a million-strong union of mainly small-scale farmers which opposed GM crops. Italian authorities duly obstructed regulatory approval of GM crops, at both the national and EU levels. The Fondazione Campagna Amica brought together civil society groups to oppose agribiotech on grounds that its development model standardizes monoculture crops, threatens biodiversity and ecosystems, and undermines distinctive products which are 'Made in Italy'.

In Austria, the government was already promoting organic agriculture as an alternative economic strategy in the early 1990s. Opponents turned GM crops into a symbolic threat to that strategy. Austrian regulators unfavourably compared potential environmental effects of GM crops to methods which use no agrochemicals (Torgersen and Seifert, 2000). Civil servants drew links between the Precautionary Principle and sustainable development. In their risk-benefit analysis of GM crops, risks were always uncertain, while benefits should promote the political aim of environmental sustainability, understood mainly as organic farming (Torgersen and Bogner, 2005).

Regulatory constraints express societal conflicts

In those various ways across European countries, agribiotech was turned into an ominous symbol of agri-industrial threats, health risks, a misguided efficiency and neoliberal globalization. Protest shared a common aim: 'stopping the technology from infiltrating the food and agricultural sectors' (Schweiger, 2001: 371). Opponents created an anti-GM alliance encompassing diverse groups, for example, small-scale farmers, civil society groups and various expert networks.

In the mid-1990s, North–South civil society networks initially opposed agribiotech for strengthening corporate domination of the agri-food system. As one strategy, they environmentalized the issue to intervene in regulatory procedures on biophysical risk criteria (Buttel, 2005). European networks stigmatized GM technofixes for evading sustainability issues, degrading biodiversity and obscuring risks through optimistic assumptions. They broadened their target to the agri-industrial monoculture system as a threat to various 'quality' agricultures, local control of food production and 'natural food'. These proposals complemented general agendas for decentralizing the economy, which can be understood as eco-localization (for example, Hines, 2000; Woodin and Lucas, 2004).

In response to public controversy, regulatory authorities tightened and broadened the environmental risk criteria, which thus became less amenable to the productivist efficiency model for which GM crops were designed. Some governments restricted or even banned GM products which already had EU approval. In 1999 the UK announced a voluntary moratorium on commercial cultivation of GM crops, given the scientific uncertainties about how herbicide sprays on herbicide-tolerant crops could affect farmland biodiversity. This move aimed to allow time for commercial-scale testing of herbicide usage and its effects, thus delaying an awkward political decision. This was eventually negative, based on results of the field trials.

In 1999 several member states jointly blocked the EU Council from approving any more GM products. Some members demanded that the EU regulatory procedure first incorporate more stringent, precautionary criteria. This procedural blockage, which became known as the de facto moratorium, expressed a policy impasse over regulatory criteria, as well as difficulties in addressing public concerns.

The EU Council's de facto moratorium stimulated a legislative change towards tighter regulatory criteria for GM products. The Commission revised the 1990 Directive to encompass a broader range of risks, for example, wider effects of herbicide usage and delayed or indirect effects (such as 'agronomic' ones). It also required that scientific uncertainty be made explicit about any 'identified risk', potentially as a basis to require commercial-stage monitoring for those specific risks. The Directive now required public consultation on such issues (EC, 2001). Together these changes potentially required greater evidence of safety before and even during commercialization, along with greater public accountability for regulatory decisions.

In parallel with the changes in EU law, the Commission sought to restart the EU-wide regulatory procedure, which had been stalled since 1999. In its view, a precautionary approach could not justify blockages or bans on the GM products under consideration. When the regulatory procedure finally resumed in 2003, however, some member states applied even more stringent criteria than in the late 1990s. Their stance generated more disagreements over new GM products, again mainly herbicide-tolerant and insect-resistant crops.

More member states challenged the available evidence of safety and raised extra uncertainties warranting extra evidence; some criticisms came from governments which anyway opposed GM crops in general (for example, Italy, Greece, Austria). Civil society groups continued to undermine safety claims, while counterposing alternative agricultures. Whenever the Commission sought EU approval for a specific GM product, there was inadequate support from member states in the EU regulatory committee, so the Commission lacked legitimacy if it approved the product. Any such decision remained vulnerable to national bans or restrictions (Levidow et al, 2005).

From commercial blockages to GMO-free zones

Civil society groups continued their demands for comprehensive GM labelling, so that consumers would not be unwittingly 'force-fed' GM food and could make their own judgements on product safety. By the late 1990s the European food industry feared consumer distrust and so voluntarily accommodated demands for more stringent labelling rules. EU regulations largely formalized the voluntary labelling rules of the food industry. This could mean a 'GM' label on any processed food.

Soya and maize were significant, pervasive components of processed food. When shipments including GM grain first arrived at European ports, wholesalers insisted that segregating non-GM grain would be unfeasibly difficult and expensive. However, the UK supermarket chain Iceland made enquiries in Brazil, gained contracts for non-GM soya shipments, established its own processing plant and then offered to sell the surplus to other supermarket chains (see Figure 3.2). Facing this new opportunity (and competitive threat from Iceland), they made arrangements to secure non-GM sources.

European food retail chains gradually excluded GM grain from their own-brand products, rather than apply a 'genetically modified' label. By

Figure 3.2: Iceland supermarket chain as David defeating Monsanto

Source: Russell Ford, Iceland Group

1999 GM grain was being blocked by this commercial boycott of the European food industry, more than by any regulatory obstacles. Lacking a market for GM grain, European farmers were deterred from cultivating GM crops, except in some maize fields in Spain, which otherwise would have a shortage of animal feed. Consequently, GM grain found a market only for animal feed, whose products required no 'GM' label (Levidow and Bijman, 2002).

The EU Council and the Parliament both requested new legislation broadening the criteria for mandatory 'GM' labelling. Formerly this requirement depended upon the presence of detectable DNA or protein in GM food. Under a new law, 'GM' labelling was now required for any food or feed product containing GM material, regardless of detectability; and its presence must be kept traceable throughout the agri-food chain (EC, 2003a, 2003b). This effectively required labelling for a broader range of products than before. With this extra weapon, civil society groups made further efforts to deter food companies from using GM grain. Soon it was limited to animal feed, which required no label.

The inadvertent spread of GM material became another contentious issue. European agribiotech opponents had warned against the prospect that GM pollen could irreversibly 'contaminate' non-GM crops of the same kind and so spread over generations. This metaphor was eventually given several meanings, both literal and metaphorical. For example, the GM material could have unknown harmful effects on recipient plants; also that corporate influence was contaminating our political system and undermining democratic accountability (Levidow, 2000). As a pollution metaphor, 'GM contamination' acquired multiple meanings and thus served as a bridge frame aligning otherwise disparate frames.

As an economic consequence of 'GM contamination', a non-GM crop may warrant a GM label, thus incurring financial loss. Such damage could be averted by GM farmers adopting segregation measures to limit the spread, but this could impose an economic burden and so constrain or deter cultivation of GM crops. So this issue potentially divided farmers.

To regain control over the 'contamination' issue, the Commission developed a 'coexistence' policy for ensuring farmers' free choice to cultivate GM, conventional or organic crops. This policy sharply distinguished between environmental issues, which were appropriate for risk regulation under the Directive, versus merely economic damage from the spread of 'safe' GM material to non-GM crops (CEC, 2003). But this key distinction was challenged, blurred and undermined, eventually by a legislative change. Again under pressure from the Parliament, the Commission agreed to amend the Deliberate Release Directive so that 'Member states may take appropriate measures to avoid the unintended presence of GMOs in other products' (EC, 2003a: 20), that is, regardless of environmental risks.

Such segregation measures were being developed by many national or regional authorities. For some, 'coexistence' policy increasingly meant segregation measures which would marginalize or preclude GM crops. In a Europe-wide charter of regional authorities, they discursively linked 'GMO-free zones' with **food sovereignty**, quality labels on food products and regional biodiversity. The charter identified GM crops as a threat to 'sustainable and organic farming and regional marketing priorities for their rural development' (FFA, 2005).

Moreover, the Assembly of European Regions (AER) proposed that coexistence rules could be based on feasibility studies examining the environmental, socio-economic and cultural impact of GM products. Areas could be designated as 'GMO free' in order to protect any added value of certified quality products such as products of origin or organic production; likewise for areas subject to mandatory biodiversity conservation (AER and FoEE, 2005). The AER's proposal generalized from stringent rules already being devised by some regional authorities.

Together all these restrictions excluded a market or even field tests for GM products. Ostensibly about 'GM contamination', this conflict expressed rival development pathways, namely: GM crops as eco-efficiency fixes for agri-industrial systems, versus quality alternatives as an agrarian-based rural development enhancing farmers' knowledge of natural resources. GMO-Free Regions conferences eventually highlighted agroecological methods as a general alternative to agri-industrial systems, thus going beyond the earlier focus on certified organic products. This support opened up greater opportunities for civil society networks promoting agroecology, as explained next.

Agroecology as a transformative agenda

From around 2010 onwards, European civil society–farmer alliances have been promoting agroecology as a general alternative. This solution diagnoses the fundamental problem as intensive monoculture systems depending on the input-supply industry, undermining farmers' knowledge of locally available natural resources and perpetuating high-carbon burdens. European agroecology alliances gained inspiration from the global South, just as the European anti-GM movement had learned from resistance there, as explained next.

Inspiration from the global South

Since the 1970s the Green Revolution has been globalizing 'technology packages' including hybrid seeds, agrochemicals and large-scale harvesting machinery. Such systems have prioritized commodity crops (for example, maize, soya and cotton), especially for animal feed and processed food. Such priorities marginalized land use for local food consumption. This capital-intensive model has likewise marginalized traditional peasant agriculture, its fertile land, its skills and market access.

Nevertheless small-scale producers still feed most people in the global South, and in some rural parts of the global North. They mostly use agroecological methods, which depend on farmers' skills and knowledge about natural resources, rather than purchased external inputs. They use less intensive cultivation attracting fewer pests and controlling them with predators and wider biodiversity. A global agroecology movement has been improving traditional methods through knowledge-exchange among small-scale producers and with agronomists (Martínez-Torres and Rosset, 2014).

Linking peasant movements globally, La Via Campesina has promoted agroecological methods as means of both 'feeding the world and cooling the planet' (LVC, 2009). As it has further argued, 'agroecology is a strategic part in the construction of food and popular sovereignty'. This means 'the right of peoples to define their own food, agriculture, livestock and fisheries systems' rather than the food supply being largely subject to international market forces (Surin Declaration, 2012; also LVC, 2013). This agenda has provided a strategic focus for political alliances: 'For peasants and family farmers and their movements, agroecology helps build autonomy from unfavourable markets and restore degraded soils; social processes and movements help bring these alternatives to scale' (Rosset and Martínez-Torres, 2012: 17).

Through the 'counter-globalization' (later 'global justice') movement, Northern activists learned from their counterparts in the global South. European civil society–farmers' networks have elaborated agroecology as a political strategy of agrarian movements over the past decade (for example, Van der Ploeg, 2009; Sevilla Guzmán and Woodgate, 2013). These European networks increasingly linked agroecology with food sovereignty (for example, Hilmi, 2012).

As a key network, the Agricultural & Rural Convention 2020 has regularly coordinated agendas for food sovereignty including agroecology. According to its early platform, 'the solution lies in a high degree of self-sufficiency and food sovereignty at local, regional, national or continental level', where people have 'the right to establish their own agriculture and food policy' (ARC2020, 2010). Or more expansively, 'Food sovereignty is the right of peoples to healthy and culturally appropriate food produced through ecologically sound and sustainable methods, and their right to define their own food and agriculture systems' (Nyéléni, 2007).

European farmer–civil society alliances have promoted agroecology as a transformative agenda. Its innovative practices have a great potential for widely replicating small-scale initiatives and relevant capacities (Ilieva and Hernandez, 2018). This alternative innovation agenda has provided a stronger basis to denounce EU policy for subsidizing resource-burdensome, capital-intensive agri-industrial systems. The EU instead should support agroecological systems for providing public goods, demanded Fridays for Future (FfF, 2020). Thus the youth climate strikes gave a higher profile to long-standing demands of civil society groups.

Agroecological methods can be outscaled for strengthening climate resilience, maintaining biodiversity and providing food security. Agroecological alternatives link more diverse food production with ecosystem services as a public good; they do so through nature based methods, circular economy principles and appropriate green technologies; the latter include traditional biodiverse seeds, soil management for pest control, nutrient recycling, and so on. Through these approaches, women and youths have gained opportunities for livelihoods, according to a research report by Agroecology Europe (AEEU, 2020a).

Crucial support has come from solidarity relationships mobilizing civil society actors and their resources. Agroecological producers have found or mobilized such support through wider networks of the solidarity economy, which seeks 'a whole basic transformation of the agricultural and food system, in which agroecology plays a key role'. Its solidaristic activities include knowledge exchange, seeds exchange and short food-supply chains bringing producers closer to consumers (AEEU, 2020b: 29). Those perspectives have guided collective demands on national and regional authorities for support measures from the CAP. By contrast, its prevalent financial incentives have driven a narrowly defined 'efficiency', seeking to increase yields of a single crop.

The agroecology-solidarity economy agenda gained support from the European Economic and Social Committee (EESC), representing civil society groups. It has demanded a shift in EU subsidy priorities. As it argues, Europe could feed its entire population by 2050 through an agroecological transformation which integrates livestock breeding, crops and trees, with a zero-carbon emissions target (EESC, 2019). This innovation potentially transforms or replaces the agri-industrial production system.

To gain better livelihoods, agroecological producers have sought to address two related difficulties – low productivity and low remuneration. Solidaristic means could address both difficulties, as shown here.

Knowledge-based productivity

The European organics movement set up Technology Platform Organics to formulate and promote research agendas favouring agroecological methods. To raise productivity, it has elaborated the novel concept 'eco-functional intensification', for linking farmers' knowledge with scientific research. It provides an alternative to the ambiguous concept 'sustainable intensification'. Eco-functional intensification means 'more efficient use of natural resources, improved nutrient recycling techniques and agro-ecological methods for enhancing diversity and the health of soils, crops and livestock. Such intensification builds on the knowledge of stakeholders using participatory methods' (Niggli et al, 2008: 34).

Eco-functional intensification is illustrated by resource conservation and recycling:

> Diversified land use can open up new possibilities for combining food production with biomass production and on-farm production of renewable energy from livestock manure, small biotopes, perennial crops and semi-natural non-cultivated areas. Semi-natural grasslands may be conserved and integrated in stockless farm operations by harvesting biomass for agro/bio-energy and recapturing nutrients from residual effluent for use as supplementary organic fertiliser on cultivated land. (Schmid et al, 2009: 26)

This vision goes beyond mixed farming, that is, arable and animal production, towards synergies with renewable energy through inter-farm cooperation. In such ways, renewable inputs and agro-biodiversity have been linked across scales to the wider landscape.

According to the EU's Standing Committee on Agricultural Research, agricultural improvements have arisen through social-experimental processes linking farmers, agronomists and citizens' groups: there are 'ongoing experiments and a re-development of knowledge networks' (SCAR FEG, 2008: ii). Agroecology should be given priority, according to its later report: 'Approaches that promise building blocks towards low-input high-output systems, integrate historical knowledge and agroecological principles that use nature's capacity and models nature's system flows, should receive the highest priority for funding' (SCAR FEG, 2011: 8). A global expert study promoted agroecological methods, especially by highlighting farmers' knowledge and innovation which lack official recognition as such (for example, IAASTD, 2009). Such proposal became extra evidence for efforts by Technology Platform Organics to shift the EU's research programme towards agroecological priorities.

In an EU policy context emphasizing innovation, mainly meaning capital-intensive technology (for example, CEC, 2010b), agroecology has been promoted as a different kind of innovative practice. It combines four types of innovation – know-how, organizational, social and technological – each type integrating farmers' knowledge (IFOAM EU Group et al, 2012). A civil society–farmer alliance has likewise advocated 'agro-ecological innovation' as an action-research project (ARC2020 et al, 2012).

Agroecology was central to a multi-stakeholder report on improving Europe's agriculture. It criticized various technofixes for reinforcing productivist management practices that harm the environment while marginalizing real solutions. Moreover, agroecological alternatives often have been fragmented: 'agroecology has been treated as a set of discrete technologies rather than as a systemic alternative' (IPES Food, 2019). Within the former, agroecology remains merely another tool-kit or technofix for the dominant agri-food system. The report advocated several policy changes for an agroecological transition to sustainable food systems (IPES Food, 2019).

EU policy conflicts: productivist versus agroecological priorities

In the EU, the CAP has been the legislative framework influencing the production, sale and processing of agricultural products. Its support prices and subsidy criteria have stimulated some markets which otherwise would not exist. These have generally incentivized technofixes, that is, external inputs for higher yields. This productivist model has generated surpluses, provided 'cheap' food for consumers, degraded natural resources, undermined biodiverse seeds and environments, and driven many small-scale farmers into that perverse model or (more often) out of agriculture.

Facing such criticism, the CAP gradually shifted the subsidy criteria away from production outputs, instead towards production units, that is, land area or livestock. Yet these reforms still incentivized the productivist paradigm. For the 2014–2020 CAP, farmer–civil society alliances proposed 'greening' measures so that subsidy would come with mandatory conditions effectively favouring agroecological practices. The European Commission entered the EU-wide negotiations with similar proposals. But these 'greening' proposals would undermine farmers' economic competitiveness, according to political lobbies representing the agri-input supply industry and mainly large-scale farmers (for example, COPA-COGECA, 2013a).

Accommodating that opposition, the European Parliament weakened the Commission's proposals, resulting in only modest improvements. Rather than be mandatory, 'green' criteria became mere options, remaining compatible with agrochemical usage and guaranteeing no biodiversity improvements. 'There is plenty of business as usual with some improvements but no real change', argued the main alliance of civil society groups (ARC2020, 2013; see also IFOAM EU Group, 2013; Pe'er et al, 2014; Lakner, 2016).

In the run-up to EU decisions on the post-2020 CAP, civil society groups again sought criteria to incentivize and remunerate agroecological practices. But this effort encountered a Right-wing political shift. In 2019 the European Parliament voted to continue subsidies for the most intensive practices and to weaken support for some measures conserving biodiversity. As a European Parliament report bluntly noted, 'The new green architecture is much more flexible in its design and management, being entrusted to the national authorities', that is, again relegating green measures to a merely optional status. Some measures such as agroecological farming became obscured or marginalized (Massot, 2019).

Despite those limitations, the CAP has included some helpful measures which can subsidize agroecological practices providing public goods. Friends of the Earth Europe has promoted policy changes in the CAP criteria to favour agroecology, at EU and national levels (FoEE, 2014). As a possible advance, the 2023–2027 CAP introduced Eco-schemes, voluntary programmes meant to incentivize more ecological and environmentally friendly farming practices; yet again, agroecological methods were relegated to mere options among other tools (AEEU, 2022). These significant changes would require a stronger alliance to overcome the capital-intensive productivist agenda of the agri-industry lobby.

Agroecological initiatives have benefited from CAP programmes in some EU member states, as positive models for other countries. A farmer–citizen campaign has sought means to improve and use the CAP for agroecology:

> This means ensuring a strong and stable support environment for high quality food production based on sustainable systems and practices such as organic farming and other environmentally-friendly farming practices with strong support for agroecological knowledge transfer, advice, cooperation and innovation. Consumers should be able to purchase ecologically-produced food from local and regional producers. Therefore the creation of short, decentralised supply chains, diversified markets based on solidarity and fair prices, community-led initiatives and closer links between producers and consumers locally and regionally must be prioritised. (ARC2020 et al, 2015: 1)

Relocalization through short supply chains

Low remuneration has been another major difficulty for agroecological producers. Likewise some organic producers have lost much of the premium price to supermarket chains, partly through competition with low-priced imports. A general remedy has been variously known as alternative agri-food networks, short food-supply chains (*circuits courts*), or agri-food relocalization.

Such arrangements avoid profit-driven intermediaries and so enhance income for producers, while also making good-quality food more accessible for low-income consumers. Forms include: direct sales, farmers' markets and public procurement, especially for school meals. A widespread form is Community Supported Agriculture, where farmers and consumer-subscribers share the risks and responsibilities. Such initiatives are necessary to incentivize and remunerate agroecological methods through consumer support, especially for farmers lacking the premium price of organic-certified products.

All these arrangements bring food producers closer to consumers in multiple senses. More generally, agroecological methods provide a basis for solidaristic relationships among producers and with consumers, bypassing competition for cheap food (ARC2020, 2022).

More ambitiously, short supply chains can empower new citizen–community alliances, as a counter-weight to the dominant agri-food system and its competitive pressures (Fernandez et al, 2013). Several initiatives have built consumer support for agricultural methods which minimize external inputs and enhance aesthetic food qualities, among other benefits. Agroecological farmers pursue production methods which can preserve the environmental quality of landscapes; they maintain agro-biodiversity by preserving local traditions and varieties. Although most initiatives started by marketing organic products, this base expanded opportunities for agroecological methods more generally to gain better remuneration. Farmers have established collective marketing initiatives in order to retain their specific product identities, proximity to consumers and the value added (Karner, 2010; Kneafsey, 2013).

Those localization initiatives build links among several activities: agroecological practices, lower external inputs and short food-supply chains. In addition to better livelihoods, participants seek to enhance their own self-determination and self-esteem. In one of many examples, Les Bons Repas de l'Agriculture Durable (BRAD), in Brittany, a citizen-led certification scheme has evaluated whole-farm sustainability. Farm visits are made by an agronomist, the first to collect data and the second to give feedback and negotiate a progress agreement with the farmer. These practices generate a group commitment to continuous agroecological improvement, rather than a priori criteria for certification (Galli and Brunori, 2013).

Within the Food and Agriculture Organization (FAO)'s agroecology programme, its 2016 European and Central Asian symposium was shaped by civil society and farmers' groups. According to their recommendations, agroecology principles should be formulated and used 'to transform and improve the current food system, based on participation, alliances and putting food producers at the centre' (FAO, 2016: 3). Alongside improvements in production methods, it called for improving short supply chains which favour small-scale producers, such as direct marketing and value adding, peasant markets, micro-dairy, Community Supported Agriculture initiatives and the Participatory Guarantee System, that is, product certification based on knowledge exchange for continuous improvement.

The ARC2020 network has highlighted transformative agroecological initiatives all over Europe. 'They value food producers, they promote the re-localisation of food systems and empower local actors' (Nyéléni Europe and Central Asia, 2021: 21). The report identified several obstacles to expansion, especially land access for small-scale farmers. Broader alliances are needed to intervene into EC policies, especially the CAP, the Farm to Fork Strategy and European Green Deal, as the report argued. Similar interventions have been elaborated for promoting food localization through short food-supply chains (ECVC, 2020).

As a multi-stakeholder organization, Agroecology Europe has advocated an agroecological transition, which needs several support measures. Short supply chains should be based on solidaristic relationships for a bottom-up governance. Measures include:

> Rethink[ing] the relation between urban and rural societies and territories, as cities and rural life and ecosystem are interdependent. We need to create an alliance based on short supply chains with surrounding territories. …
>
> The transition to healthy systems must be done by giving a voice to the ones that are never heard. Bottom-up governance mechanisms for the European food systems including farmers, food workers, citizens and social movements as fundamental agents are needed. Transitioning means moving from food production to food security and sovereignty, through participatory governance. (AEEU, 2021)

Together those activities build a knowledge commons, maintained by solidaristic communities across diverse spaces. In such ways, the transformative potential of agroecology depends on integrating its three forms – transdisciplinary knowledges, interdisciplinary agricultural practices and social movements – while recognizing their mutual dependence (Wezel et al, 2009). This potential likewise depends on integrating knowledge-exchange, grassroots innovation and an eco-localization agenda. Together all this strengthens solidaristic alternatives to the dominant agri-industrial system.

Climate-resilient agriculture: rival agendas

As climate change became a more prominent issue, there were various proposals for agricultural systems to strengthen their resilience vis-à-vis climate stresses, as well as to reduce their GHG emissions. Long-time disagreements arose over systemic sources of such climate problems. Driven by global commodity markets, intensive monoculture methods attract pests and try to control them through agrichemicals, whose production entails significant GHG emissions. As a more complex issue, soil tillage offers many benefits such as weed control, manure incorporation and seedbed preparation. But tillage can cause several problems, for example, soil erosion, GHG emissions, fertility loss and run-off including agrichemicals; crops may become more vulnerable to stresses. How to reconcile agronomic, climate and other environmental benefits?

Since the turn of the century, agribiotech companies were promoting GM herbicide-tolerant crops with herbicide sprays as a no-till method to control weeds. This method eventually became central to a more ambitious agenda of 'climate-smart agriculture' (CropLife, 2014). Yet the concept had different meanings based on agroecology (FAO, 2013). Rival models again contended for influence.

At the September 2014 UN Climate Summit, the FAO and World Bank jointly launched the Global Alliance for Climate-Smart Agriculture (GACSA). This agenda was promoted by some nature conservation organizations, businesses and numerous governments, especially France as the EU Presidency. GACSA featured multinational companies prominent in GM crops and pesticides (Monsanto and Syngenta) as well as Yara, the world leader in synthetic fertilizer. Such capital-intensive inputs were claimed to bring three main benefits: 'sustainable and equitable increases in agricultural productivity and incomes; greater resilience of food systems and farming livelihoods; and reduction and/or removal of greenhouse gas emissions associated with agriculture' (GACSA, 2014).

GACSA's agenda was soon denounced as a 'false solution' and 'corporate-smart greenwash'. Such criticisms came initially from social movements, soon joined by mainstream NGOs such as Greenpeace, Action Aid and Oxfam. In particular, they argued that 'carbon sequestration in soils is not permanent and is easily reversible, so it should be excluded from schemes to offset GHG emissions' (Action Aid Intl, 2014).

From their critical perspective, this harmful agenda 'welcomes industrial approaches that drive deforestation, increase synthetic fertiliser use, intensify livestock production or increase the vulnerability of farmers'. They warned about the prospect of unequal power relations, whereby 'the agendas of corporations and wealthy governments are given greater weight than those of civil society organisations, small-scale farmers and developing countries' (CSA Concerns, 2014). In that regard, GACSA refused to define criteria for stakeholder engagement, nor establish accountability mechanisms, as key criticisms from Catholic development agencies (CIDSE, 2015).

Some critics argued that the real objective is to accelerate 'the industrialisation and financialisation of agriculture'. This agenda would undermine agroecological peasant agriculture, which provides the real solution: 'We can feed people while cooling the climate', declared the European affiliate of La Via Campesina (ECVC, 2015).

As a potential basis for carbon credits, moreover, carbon sinks would enable subsidies for unsustainable practices such as tree plantations and might award credits to carbon storage which turns out to be unstable. Therefore 'mitigation in the land-use sector should not be used as offsets to displace or reduce mitigation in other sectors' (Carbon Market Watch, 2014). If reforestation of depleted forests accrues carbon credits, moreover, then soil could be redefined as a giant carbon sink to offset perpetual GHG emissions, in turn as basis to revive markets in carbon credits. Even worse, this could incentivize land grabs for financial interests (Paul, 2015). According to such critics, agroecological methods offer a more reliable means to reduce GHG emissions from agriculture and to enhance its resilience (Pimbert, 2017; also GRAIN, 2016).

The climate-smart GM-fix soon lost support and enthusiasm, for several reasons. Given the persistently low carbon price, carbon credits were becoming less attractive and thus likewise 'carbon sinks' as a basis. According to greater scientific knowledge about soil, long-term carbon storage there would be difficult or unreliable; for example, as plant growth increases, soil carbon does not. Moreover, as rising atmospheric CO_2 levels stimulate plant growth, soil carbon levels could plausibly decline. As climate change worsened, much land was becoming a carbon-emissions source rather than a sink (van Groenigen et al, 2014; Schimel et al, 2015; Terrer et al, 2021).

Having already supported agroecological alternatives, European Green Parties denounced 'climate-smart' approaches for misdirecting agriculture to sequester carbon or to compensate for GHGs released through agri-industrial production. This would be an irresponsible 'climate sale of indulgences', by analogy with the medieval Roman Catholic Church receiving payments to absolve sins. As a contentious example, the 'climate-smart' agenda promotes herbicide-tolerant crops with no-till methods to avoid soil compaction and so conserve it. On the contrary, no-till methods could bring environmental benefits only in a biodiverse agroecological system where 'different roots from mixed crops build pores and thus protect against soil compaction' (Greens/EFA, 2020: 1, 15).

The several benefits depend on biodiverse agroecological methods rather than the agri-industrial ones of the GACSA lobby, argued its critics. As a techno-market fix, then, 'climate-smart agriculture' was soon turned into a public controversy. Counter-publics created and used the opportunity to promote agroecology as the truly climate-resilient form of agriculture. Its expansion depends on social movements advocating connections between agroecology, the right to food, food sovereignty and environmental integrity (GCA, 2019: 4–5). As a role-reversal, social movements eventually appropriated 'climate-smart agriculture' for an agroecology agenda, framing a Sustainable Agriculture Standard for responsible supply chains (Rainforest Alliance, 2020).

Conclusion

Let us return to the original questions:

- How did the EU promote agribiotech through institutional change and beneficent future visions?

- How did protest stimulate public controversy and eventually block agribiotech in the EU?
- How did opponents create and use opportunities for 'quality' alternatives?
- How do these divergent agendas relate to system change versus continuity?

To answer those questions, this chapter analysed institutional commitments and techno-optimistic assumptions underlying the EU's agribiotech fix. In particular, the policy framework promoted several market-type drivers as conditions that were supposedly necessary for agribiotech to bring the EU societal benefits.

The EU's agribiotech fix was promoted in several stages since the 1990s. Back then the EU elaborated a techno-market framework for a Life Sciences agenda: market incentives and competition were meant to generate eco-efficient GM products enhancing resource efficiency, environmental protection and economic competitiveness. More than simply a false solution, this beneficent vision served as an imperative for neoliberal policies – for example, extending patent rights, enclosing seeds-knowledge commons, making public-sector research more dependent upon private finance, directing such research to favour the input supply industry, pushing farmers into more intense market competition, and conceptually homogenizing diverse environments in order to identify no environmental risks.

From an STS co-production perspective, the agribiotech fix co-produced distinctive forms of nature (knowledge substituting genomic-biological inputs for agrichemicals, precise improvements of deficient crops, homogeneous environments as a regulatory baseline), of technoscience ('clean' genetic inserts as biotechnological inventions) and of social order (beneficent market competition; publics reduced to consumers). An EU state–industry partnership co-produced these forms through a Life Sciences sociotechnical imaginary.

This neoliberal framework made agribiotech politically vulnerable to opposition from many critical standpoints, for example, environmental risk, genetic resources, farmers' knowledge, consumer choice, public-sector research, and so on. Through mass protest from the late 1990s onwards, GM products were popularly stigmatized as a threat to the environment, sustainable development and democratic accountability. Critics diagnosed agri-industrial efficiency as a threat, drawing analogies to the 'mad cow' pandemic; thus they gave ominous meanings to the eco-efficiency claims for agribiotech. This ominous analogy served as a bridging frame (cf Snow et al, 1986), aligning the diverse frames of various stakeholder groups.

Opposition networks linked small-scale farmers, civil society groups and critical experts, jointly acting as counter-publics. They environmentalized the issues for intervening in regulatory procedures (Buttel, 2005). Pro-agribiotech forces had sought to scientize the issues as technical-expert ones, yet their expertise was contested as politically partisan. Facing widespread protest, the EU policy system shifted towards more stringent biophysical criteria for risk assessment. In particular, the 2001 revised Directive mandated a broader assessment of long-term and indirect effects, such as changes in herbicide usage related to GM crops.

As 'GM contamination' became a more prominent issue, it too became a bridge frame linking diverse critical frames such as health risks, environment risks, the state's democratic unaccountability, corporate power over the agri-food chain and non-GM farmers' lost livelihoods. The latter issue was translated into biophysical and economic effects: the EU sought regulatory criteria facilitating the 'coexistence' of GM crops with nearby non-GM crops. But some state authorities imposed segregation requirements that effectively impeded GM crops. These restrictions were eventually accepted by new EU guidance. In such ways, the EU flexibly adapted neo-regulation to accommodate or soften societal conflicts over agribiotech (cf McMichael, 2005; Busch, 2010). Indeed, at stake was the political legitimacy of regulatory decisions.

As counter-publics, civil society–expert networks used the opportunity to complicate the regulatory procedures and to deter or block market access for GM products; these efforts complemented regulatory barriers of some national and regional authorities. European networks broadened their target to the agri-industrial monoculture system as a threat to various 'quality' agricultures, symbolized by various speciality or territorial foods. The widespread slogan 'GMO-free' was extended to a GMO-Free Regions network, promoting alternative development agendas.

Using those opportunities, civil society–farmer networks soon promoted agroecological methods and consumer support through short food-supply chains. Agroecological producers thereby bypass profit-driven middlemen, educate consumers about sustainable production methods and gain better remuneration. Through knowledge-exchange for grassroots innovation, practitioners have developed production methods that can be readily adapted and replicated by others. To promote agroecological methods, campaigns have advocated specific measures for the CAP to strengthen farmers' remuneration, land access, collective capacities, market opportunities and knowledge-exchange. Although endorsed by the European Commission, such measures have been blocked or softened by the European Parliament, accommodating the agri-industry lobby agenda to maximize productivity for global market competitiveness.

Nevertheless European agroecology networks have built a commoning process which helps resist enclosures of common resources by the agri-industrial system. From a co-production perspective, the agroecology agenda has co-produced distinctive forms of nature (agroecosystems for optimizing ecological interactions), technoscience (farmers' knowledge commons about biodiverse natural resources) and social order (solidaristic relations among producers and with consumer-citizens). Through knowledge-exchange for grassroots innovation, practitioners have developed production methods that can be readily adapted and replicated by others. These sociotechnical forms facilitate eco-localization, conflicting with the techno-market fix of the agribiotech agenda.

As climate change became a more salient issue, this stimulated greater debate over agri-innovation as means to reduce GHG emissions and deal with climatic stresses. With EU support, in 2014–2015 a state–corporate alliance promoted 'climate-smart agriculture' featuring capital-intensive inputs including GM herbicide-tolerant crops with herbicide sprays as a no-till method. Moreover, land could become a carbon sink generating tradeable carbon credits, thus potentially a techno-market fix. Civil society networks denounced this agenda as a 'false solution' and 'corporate-smart greenwash'. They counterposed agroecology as the truly climate-resilient agriculture by 'feeding the people and cooling the planet'.

At both stages of the agribiotech agenda, counter-publics turned a techno-market fix into a public controversy. This became an opportunity for counterposing agroecology to the entire agri-industrial system, as a potential step towards system change. This conflict has expressed rival sociotechnical imaginaries: a Life Sciences techno-market fix versus agroecology-based eco-localization (Table 3.1).

4

EU Biofuels Fix: Prioritizing an Investment Climate

Introduction

From 2007 onwards there was a global debate on biofuels as a potential means to reduce GHG emissions from transport fuel. According to proponents, plant-based feedstock was renewable, low-carbon and thus an environmentally sustainable substitute for fossil fuels; moreover, such flexibly sourced fuels would enhance energy security. On such grounds, new government rules or proposals required a minimum percentage of biofuels in transport fuels, for example, in Brazil, the US and EU.

This agenda drew criticism and became controversial for several reasons. Claims for environmental benefits made several optimistic assumptions about feedstock production, especially oilseeds and soya. As food prices spiked, critics highlighted a conflict between 'food versus fuel' in priorities for using land and natural resources, especially water. Oilseed production for other uses had previously incentivized land-grabs, deforestation and more chemical-intensive agri-production; so producing more oilseeds for biofuel feedstock would aggravate such pressures. By stimulating such changes in land use, biofuel substitution for oil may not save GHG emissions and may even increase them, argued some experts (Searchinger et al, 2008).

Moreover, biofuel expansion was blamed for socio-environmental injustices such as resource degradation and land grabs, associated with changes in land use (FIAN International, 2008; ICHRP, 2008). In Brazilian sugarcane production for sugar or bioethanol, workers were being trapped in quasi-slave labour conditions, officially known as *trabalho escravo*. Many had to be rescued by a government agency (Mendonça, 2006, 2010; MTE, 2010). These systemic harms were later substantiated by academic and expert studies (for example, Borras et al, 2010, 2012; CETRI, 2010; Lehtonen, 2011; Matondi et al, 2011; Action Aid, 2012; Schulze, 2012). In particular, land grabs were driven by 'demand for biofuel feedstocks as a reflection of policies and mandates in key consuming countries', according to a World Bank report (Deininger and Byerlee, 2010: 11).

Beyond direct changes in land use, global trade was stimulating crop substitution and thus indirect land-use change (ILUC). As a high-profile example, the US government subsidy for bioethanol led many farmers to switch crops to maize from soya, in turning stimulating greater soya cultivation in Brazil and rainforest destruction for land-clearing there (Laurance, 2007). As the EU's leading biofuel user, Germany was importing Europe-wide sources of oilseed rape, whose former food uses continued by importing more palm oil from Indonesia, where new plantations often destroyed forests (FoEE, 2010b; IFPRI, 2010). In such ways, biofuel production was displacing crop cultivation to other places, where an initial destruction of forest or peatland generates enormous GHG emissions. This one-time 'carbon debt' undermined GHG savings from biofuels replacing fossil fuels (Fargione et al, 2008; Searchinger, 2008).

Those global criticisms generated controversy over the EU's biofuel mandate, which illustrates a techno-market fix for climate change. Under the 2009 EC Renewable Energy Directive (RED), 10 per cent of all transport fuel had to come from renewable sources by 2020; eventually a higher target was set for 2030. The initial target stimulated a biofuels market that otherwise would not exist. The EU had insufficient domestic feedstock to fulfil the target, so this would depend on greater imports, especially from the global South.

The EU biofuels agenda promised several benefits, namely: environmentally sustainable feedstock, lower GHG emissions, less dependence on fossil fuels, greater energy security and rural development wherever biofuels are produced. Those beneficent promises provoked a wide-ranging public controversy. As a response from biofuel proponents, any potential harm could be avoided through sustainability criteria and eventually through 'advanced' biofuels using cellulosic biomass rather than edible sources.

As a precursor of this chapter, Participatory Action Research (PAR) brought together academics with civil society groups questioning the implicit politics of the EU biofuel mandate. Early discussions agreed to investigate its assumptions, while comparing them with real-world practices and experiences, especially in the global South. Going beyond disputes over biophysical effects, this approach analysed political-economic drivers, institutional change, societal conflict and oppositional strategies (Franco et al, 2010; Levidow and Oreszczyn, 2012).

Extending that original inquiry, this chapter will answer the following questions:

• How did the EU justify and create a new market for biofuels? Through what institutional changes and beneficent societal vision?

- How did the EU biofuels mandate provoke opposition, controversy and alternative agendas? With what outcomes?
- How do these divergent agendas relate to system change versus continuity?

To answer these questions: The EU mandate has been analysed here as a techno-market framework, combining ecological modernization with neoliberal environmentalism (see again Chapter 2). The framework has had several institutional drivers, namely: to make the EU a buyers' market for competing feedstock suppliers; to finance R&D for future biofuels which could gain intellectual property; to strengthen the 'investment climate' for companies funding techno-innovations for a Knowledge-Based Bio-Economy (KBBE); and to normalize the rising transport flows within the EU's internal market. For some political forces advocating the EU mandate, those roles were aims, taking priority over GHG reductions. The ensuing controversy featured rival agendas, which can be understood as divergent sociotechnical imaginaries.

This chapter has the following structure: the first section on how the EU mandate served wider bioeconomy strategies to plunder global resources, while making techno-optimistic promises about societal benefits; the second section on how critics generated a controversy over EU targets as a multiple threat to GHG savings, natural resources and livelihoods in the global South; the third section on how the EU's mandatory targets included sustainability criteria somewhat accommodating criticisms, but deferring any regulatory criteria for the most harmful effects from ILUC; the fourth section 4 on how the ILUC issue was channelled into expert procedures, thus protecting an 'investment climate' for the EU's bioeconomy agenda; and, finally, the concluding section returns to the questions posed earlier. Table 4.1 summarizes the rival agendas.

Promoting EU biofuel expansion as a techno-market fix

The 2009 EC RED mandated that transport fuel must include a rising minimum percentage of renewable energy. According to its advocates, market competition to fulfil the mandate would incentivize R&D towards 'advanced' biofuels, also known as second-generation biofuels. Such innovation would contribute to a wider KBBE, bringing the EU several benefits, especially energy security, environmental protection, technology export and rural development.

Without biofuels, the EU would become more dependent on fossil fuel imports, face energy insecurity and increase GHG emissions, according to proponents. Hence 'there is a particular need for greenhouse gas savings in transport because its annual emissions are expected to grow by 77 million

Table 4.1: EU biofuels mandate: divergent sociotechnical imaginaries

Sociotechnical imaginary Issue	Techno–market fix	Critiques and alternatives
Co-producing: Nature/resources Technoscience Social order	Decomposable and recomposable biomass as inputs for global value chains. Molecular-level databases to identify, select and alter valuable biomass (El Dorado). Market competition for an advantage in global value chains.	Agroecosystems facilitating agroecological production methods. Farmers' techniques for improving agroecological methods and enhancing productivity. Farmers' solidaristic relationships including knowledge-exchange processes.
KBBE	Capital-intensive innovation will bring eco-efficient means to identify and process biomass as a substitute for fossil fuels. This substitution will reconcile environmental sustainability with economic advantage.	Agroecological production methods intensify interactions among biodiverse components of an agroecosystem, helping to enhance productivity, soil fertility, nutrient recycling and crop protection.
Conventional biofuels: GHG savings	Biofuels based on biomass from anywhere can save GHG emissions by replacing fossil fuel in an expanding transport sector. The RED assigns some emissions to co-products (for example, usable for feed or electricity), so the fuel could have lower or no emissions. Sustainability criteria can clarify and ensure GHG savings.	Feedstock should come from biowaste, not food crops. The EU mandate stimulates market demand for 'agrofuels' (a bridging frame), industrially produced, environmentally unsustainable sources. This market 'drives to destruction': Harmful land-use changes in the global South jeopardize environmental benefits of biofuels.
Second-generation (2G) or advanced biofuels	2G biofuels will offer at least two improvements: They can better save GHG emissions by more efficiently converting plant material to energy. And they can avoid competition with food by using non-food crops and/ or being grown on marginal land. A bioeconomy will turn agriculture into 'oil wells of the 21st century'.	Regardless of 2G biofuels, large tree plantations will increase land and water usage, competing with food production. Harvesting whole plants reduces soil organic matter, harming ecosystems. High-volume production will favour large monoculture plantations. An environmentally sustainable future must go beyond the internal combustion engine.

Table 4.1: EU biofuels mandate: divergent sociotechnical imaginaries (continued)

Sociotechnical imaginary / Issue	Techno–market fix	Critiques and alternatives
ILUC	Although ILUC could undermine GHG savings, this potential effect will be managed by limiting the percentage of feedstock from food crops with high ILUC-risk, whenever such effects can be reliably modelled and predicted.	ILUC effects could undermine the environmental viability of biofuels by counteracting any GHG savings from oil substitution. ILUC's initial carbon debt creates a long-term time-bomb which cannot be reliably managed.
Degraded, semi-arid or marginal land	By cultivating biofuels feedstock, degraded or semi-arid land could be put back under vegetation cover by planting species adapted to such adverse conditions. Future novel crops could use 'marginal land' which lacks any other use.	Substantial land is mis-characterized as 'marginal' to justify its appropriation from people who depend on its resources, thus marginalizing them. In practice, commercial investors seek better-quality land linked with infrastructure.
Rural development	Biofuel markets spur rural development by creating new jobs, diversifying incomes and thus enhancing livelihoods, in both the global North and global South, including countries of greatest rural poverty.	Mandatory markets for biofuels shift land use to agri-industrial production of commodity crops. This favours powerful economic actors who can make such investments, thus jeopardizing local food production and efforts at food sovereignty.
Fuel security	For biofuels, feedstock can come from diverse biomass types, increase competition among suppliers, offer more flexible supply chains, give buyers greater control over markets and thus enhance the EU's energy security. A larger, stable supply of liquid fuel is needed for European competitiveness.	The 'energy security' goal complements the Global Europe strategy for EU-based companies to gain greater access to the world's natural resources within and beyond the EU (CEO, 2008). The RED perpetuates this drive to gain cheaper resources from the global South.
EU Fuel-efficiency standards	Biofuels will save GHG emissions, justifying delays in more stringent long-term standards for fuel efficiency.	Claims about biofuels saving GHG emissions serve as a pretext to avoid more stringent EU standards for fuel efficiency and thus GHG emissions.

Note: Underlying the EU's original biofuels mandate was a techno–market fix, promising several future benefits. This policy framework provoked criticism from various stakeholder standpoints. But they did not coalesce into a single opposition, nor a coherent sociotechnical imaginary, much less a mass campaign.

tonnes between 2005 and 2020 – three times as much as any other sector'. Consequently, 'the only practical means' to gain energy security is biofuels, along with efficiency measures in transport, argued the European Commission (CEC, 2007a).

This circular rationale normalized the EU's rising demand for transport fuel, which had several drivers: the EU project to 'complete the internal market', its pressures for more intense competition and EU subsidy for transport infrastructure (Bowers, 1993; Fairlie, 1993). Since the 1990s multinational companies had jointly lobbied for faster, more extensive and more efficient transport links. As the rationale, such links were necessary 'to better enable European companies to respond to the rapid and time-sensitive delivery of goods caused by globalisation and growth in world trade' (Business Europe, 2009). Supposedly responding to such market pressures, EU policies intensified them.

Moreover, greater GHG emissions were facilitated by lax standards for fuel efficiency, increasingly citing future biofuels as an alibi (Green Car Congress, 2006; T&E, 2006). When the European Commission proposed more stringent mandatory standards, especially 120g CO_2/km as an average for each company's new sales, this proposal was weakened by industry lobbying (T&E, 2009). In particular, German manufacturers complained that such a standard would threaten their EU-wide market for luxury cars. The industry successfully argued that GHG emissions would be reduced more effectively through the parallel mandate for 10 per cent renewable energy in transport fuel by 2020 (CEO, 2007, 2009). This strong lobby was backed by the German government, among others. Consequently, EU Regulation 443/2009 set an average CO_2 emissions target for new passenger cars of 130g/km – little more stringent than before.

For the subsequent decade, technological promises for future biofuels provided an extra alibi to normalize GHG emissions from EU transport and its fuel inefficiency. In 2012 the EU Cars 2020 Action Plan had weaker, slower targets than in the Commission's original proposal (T&E, 2012a). When it again proposed more stringent standards, these were blocked by Germany with UK support (Carrington, 2013). 'It is outrageous that EU countries, prompted by Merkel's government, bent to the interests of luxury German carmakers at the expense of the environment, jobs and the wider economy', declared an environmental group (T&E, 2013).

The EU biofuel mandate was officially justified as necessary for climate protection, yet a stronger impetus was the aim for greater corporate control over global resources. This aim was expressed as the EU's competitive advantage in several senses: Relative to oil, agricultural commodities offer buyers more flexible supply chains and thus greater control over feedstock: 'Biofuels add to energy security by increasing the diversity of fuel types and of regions of origin of fuels. It is not obvious how to place a monetary value on this benefit' (CEC, 2007b: 10, 12; also Barroso, 2007).

Through a KBBE, moreover, capital-intensive innovation would bring eco-efficient means to use biomass feedstock as a substitute for fossil fuels, thus reconciling the EU's environmental sustainability and economic advantage. This sociotechnical imaginary featured an economic imaginary of Europe as a global competitive space; biofuels would provide both an EU instrument and beneficiary of global markets. Environmental sustainability would result from eco-efficient productivity through resources which are renewable, reproducible and therefore sustainable. Renewable raw materials would provide biomass for flexible conversion into non-food products, especially energy and other industrial products (DG Research, 2005).

These societal visions of a KBBE had been elaborated by European Technology Platforms, representing major multinational companies in the agri-food-forestry-biofuel sectors. Market-industrial metaphors were projected onto natural resources; for example, agriculture will be 'oil wells of the 21st century'. By 2030 'Integrated biorefineries co-producing chemicals, biofuels and other forms of energy will be in full operation' (Biofrac, 2006: 16). Technoscientific advance will turn bioresources into an 'El Dorado'. By analogy with crude oil, 'biocrude' components of plant cells will be 'cracked' into their components, thus reconstructing natural resources as industrial raw materials (Levidow et al, 2013). Such R&D agendas prioritize capital-intensive techniques which can gain intellectual property (Birch et al, 2010).

This beneficent future was symbolized by the promise of 'second-generation' biofuels from non-edible cellulosic biomass. This agenda was supported by the most powerful farmers' organization, COPA-COGECA, in alliance with the European Technology Platforms. Complementing their role, the European Commission's DG Agriculture promoted both conventional and second-generation biofuels as a long-term income source for European farmers (DG Agri, 2007). Indeed, a guaranteed market was anticipated as a farm subsidy.

This overall sociotechnical imaginary was incorporated into the Commission's policy framework: 'By actively embracing the global trend towards biofuels and by ensuring their sustainable production, the EU can exploit and export its experience and knowledge, while engaging in research to ensure that we remain in the vanguard of technical developments' (CEC, 2006a: 5). In parallel, 'long-term market-based policy mechanisms could help achieve economies of scale and stimulate investment in "second generation" technologies which could be more cost effective' (CEC, 2006b). They were expected to 'boost innovation and maintain Europe's competitive position in the renewable energy sector' (CEC, 2007b). Hence the EU's biofuels mandate served as a techno-market fix, depending on techno-optimistic assumptions and displacing the state's responsibility for GHG reductions.

Opposing 'agrofuels' as a multiple threat

Early on, critics pejoratively renamed agri-industrial biofuels as 'agrofuels'. They denounced the EU for citing climate objectives as a pretext. According to an NGO, the Corporate Europe Observatory:

> The Commission's agrofuel policy has not been driven by the fight against climate change. It has sought to secure energy supply and serve the needs of large farmers and agribusiness, alongside the automotive, oil and biotech sectors, all with a direct interest in maintaining the existing status quo. The Commission has enabled these corporate interests to enter into the policy dialogue and design policy outcomes by setting up advisory groups with a clear industry bias. (CEO, 2007)

This policy framework was driven by multinational companies seeking to gain an advantage from more intense competitive pressures within and beyond the EU, argued critics (CEO et al, 2007; CEO, 2008).

By 2007 greater land use for biofuel expansion provoked controversy, questioning claims for societal benefits. Several Latin American organizations published an Open Letter to EU institutions and citizens, saying that 'We Want Food Sovereignty Not Biofuels'. They opposed biofuel expansion as a threat to their environments and livelihoods (WRM, 2007). The Open Letter was soon followed by a similar declaration from over 200 NGOs opposing EU incentives for agrofuels (Econexus, 2007). In parallel, large NGOs such as Greenpeace and Friends of the Earth Europe focused on the EU's palm oil imports from southeast Asian plantations for destroying rainforests and wildlife habitats.

More fundamentally, some development NGOs denounced the EU's agrofuel agenda for further stimulating agri-industrial methods which threaten natural resources and community access to them. The agrofuel threat comes from 'the intensive, industrial way it is produced, generally as monocultures, often covering thousands of hectares, most often in the global South'. EU targets were already stimulating land grabs displacing traditional usages, in anticipation of larger biofuel markets (Econexus et al, 2007: 6, 24).

Their harmful effects were documented in numerous NGO reports, informed especially by North–South activist networks focusing on rural development issues (ABN, 2007; Econexus et al, 2007; WRM, 2007; Bailey, 2008; Eide, 2008; Nyari, 2008). European networks attacked EU policy agendas which increase demand for transport fuel, perpetuate oil addiction and seek a false solution through biofuels (for example, ASEED, 2008).

By stimulating changes in land use, moreover, biofuel substitution for oil did not guarantee GHG savings and may even increase GHG emissions (Searchinger et al, 2008). Biofuels also faced criticism for unsustainably using natural resources, especially by comparison with alternative options. Biomass conversion into combined heat and power offers greater efficiency and GHG savings than biofuels, according to many expert reports (for example, SRU, 2007). Indeed, there are better ways to achieve GHG savings and strengthen secure supply than to produce biofuels. 'And there are better uses for biomass in many cases', argued the EU's Joint Research Centre (JRC, 2008: 22).

Given all those contentious issues, the 2007–2008 global biofuel controversy potentially undermined the EU's plans for a RED. Underlying its beneficent techno-optimistic assumptions, the EU's sociotechnical imaginary was enacted at many levels, potentially depoliticizing the issue. The EU's global resource plunder was greenwashed through narrow sustainability criteria and eco-efficiency promises (see the next section).

Central to the dominant sociotechnical imaginary, eco-efficient conversion techniques would spare good agricultural land on which local people depend. According to the EU Trade Commissioner, who was also promoting trade liberalization: 'We have all seen the maps showing the vast tracts of land that would be required to replace petrol to any significant degree. That is why research and development into second generation biofuels that are cleaner, more versatile, and can be used on more marginal land is so important' (Mandelson, 2007).

Enacting this imaginary, the EU's Framework Programme 7 allocated substantial funds to research future biofuels and biorefineries. A high priority was liquid fuel that could fulfil the target in the medium term. Such R&D priorities were driven especially by the European Biofuels Technology Platform, representing the energy, motorcar, chemical, biotech and other industrial sectors (EBTP, 2008; see protest in Figure 4.1).

The Framework Programme sought a horizontal integration across those sectors through a biorefinery that could convert diverse feedstocks into higher added-value products as well as energy (EBTP, 2008: SRA-23). Such innovations were also expected to use 'marginal land' for growing novel non-food crops and to turn 'bio-waste' into energy (for example, EBTP, 2008: SRA-24; cf Mandelson, 2007). Such spaces were seen as 'under-utilized' or 'under-valued', thus rendering invisible their importance for local livelihoods and resources.

Such technological-managerial solutions assumed that inefficient resource usage causes the sustainability problems of biofuels. This assumption is contradicted by the history of technoscientific development, especially in agriculture, where new production techniques have stimulated greater plunder of resources. Similar incentives would arise from novel techniques

Figure 4.1: At the 2008 conference of the European Biofuels Technology Platform, the protest slogan says, 'Agrofuel – no cure for oil addiction and climate change'

Source: ASEED

which more efficiently convert biomass into valuable products (Smith, 2010: 120; Levidow and Paul, 2011).

Stimulating 'sustainable' biofuels and investment

Facing the global controversy, the European Commission sought more authoritative, publicly credible criteria for 'sustainable biofuels' that would be eligible for the targets (Euractiv, 2008). The biofuel controversy was translated into sustainability criteria selectively accounting for carbon, in turn justifying mandatory EU targets, as shown in this section.

In January 2008 the Commission published its proposal for a Climate and Energy Package, including a draft Directive promoting 'the use of energy from renewable sources'. This mandated targets for transport fuel to include 'renewable energy', thus downplaying biofuels per se. That target provoked much dissent, even among EU institutions, especially from DG Environment and the Joint Research Centre:

> We raised criticisms but were ignored. This is not 'robust science policymaking' that considers all the evidence. Instead the policy is driven by sound-byte science to support a particular viewpoint. ... In the renewable energy sector, the policy picks winners – a specific

industry for biofuels. (interview, Senior Researcher, Joint Research Centre, 11 March 2010)

This process was likewise criticized as 'policy-based evidence gathering', that is, a process whereby evidence is selected to support a previously determined policy (Sharman and Holmes, 2010).

To ensure environmental sustainability of the mandatory target, second-generation biofuels were originally meant to be a pre-condition. At its March 2007 meeting the EU Council supported mandatory biofuel targets reaching 10 per cent in 2020 – subject to production being sustainable and second-generation biofuels becoming commercially available. However, the Commission's legislative proposal rejected such a condition for undermining industry investment: 'The main purpose of binding targets is to provide certainty for investors. Deferring a decision about whether a target is binding until a future event takes place is thus not appropriate' (CEC, 2008b: 17). Thus future biofuels were demoted to a rationale for investment incentives and thus an alibi for conventional biofuels.

Conflicting interests and aims sought to shape the legislative mandate for renewable energy in transport fuel. EU agricultural interests generally supported high targets, alongside sustainability criteria that would effectively limit imports from the global South. Business interests jointly sought 'a high and mandatory biofuels target, providing them with the long-term visibility necessary for their investments' (Dontenville, 2009: 39, 54).

Parliamentary rapporteurs for the draft Directive had proposed to raise the general requirement for GHG savings, for example, in order to limit harmful effects of indirect land use change (Corbey, 2007; Turmes, 2008). For several years Green Members of the European Parliament (MEPs) had been advocating biofuels from biowaste materials, which now were being marginalized by feedstock from food crops; such politicians now criticized the expansion of 'agrofuels' (Lipietz, 2008). Nevertheless Green MEPs generally supported the 10 per cent target with proposals for mandatory sustainability criteria.

The year before the 2009 Directive was enacted, the proposal was criticized by 17 NGOs, including large environmental organizations. According to their joint statement, the EU's mandatory biofuel target must be suspended unless it added substantial environmental and social safeguards; these must encompass issues such as carbon sinks, ecosystems, indirect effects and social standards (T&E, 2008). North–South NGOs soon persuaded large environmental NGOs to harden their stance against the 10 per cent target: they jointly asked the EU 'to withdraw proposals to massively expand the use of biofuels' (FoEE et al, 2008; see also CEO et al, 2008).

Despite those widespread criticisms, the 2009 RED had a clear priority: 'mandatory targets should provide the business community with the long-term stability it needs to make rational, sustainable investments in the renewable energy sector' (EC, 2009: 16, 17; cf CEC, 2008b). As the main incentive, 20 per cent of all energy had to come from renewable sources by the year 2020; also 10 per cent of all road and rail transport fuel must come from renewable energy by then. Sustainability criteria defined which biofuels qualify for the targets: GHG savings must exceed 35 per cent; this requirement rose to 60 per cent for new biofuel installations in 2017.

Relevant to such calculations, the RED double-counted GHG savings for several categories: wastes and residues, assuming that they have no other use; 'advanced biofuels' from non-edible material; and co-products which could be used for other energy sources or animal feed. For the latter bonus, economic operators use the 'energy allocation' method, whose calculation 'ends at the factory door'; beneficiaries need not demonstrate that co-products substitute for feed in practice (interview, DG Energy, 8 July 2009). At the same time, the bonus system made a beneficent assumption that co-products do substitute for production elsewhere and thus further save GHG emissions (EC, 2009: 25; also CEC, 2010c: 13). Together those criteria were meant to stimulate biofuels generating more co-products and novel biofuels more efficiently converting non-edible biomass.

The RED specified adverse changes in land use which would exclude biofuels from eligibility for the target. Environmental criteria disqualify any sources from 'highly biodiverse', 'primary forest' and 'continuously forested' areas; the latter were defined by statistical criteria. Producers should avoid 'the conversion of high-carbon-stock land that would prove to be ineligible for producing raw materials for biofuels and bioliquids' (EC, 2009: 24). Compliance would be assessed on the basis of company information, or through voluntary certification schemes or bilateral and multilateral agreements.

Within its biofuel policy, the Commission imagined market drivers benefiting rural populations in the global South. Namely, new biofuel markets would offer increased productivity, more profitable and diversified agricultural sectors, value-adding industries in rural areas, more rural employment and less migration to urban centres (CEC, 2008a). Rural populations would be incorporated into biofuel development processes as labourers in large-scale monocrop biofuel production processes; in parallel, smallholders can engage in contract-growing schemes, according to the Commission's agency for developing countries (EuropeAid, 2009). The policy framework saw rural development primarily as higher individual incomes from waged labour or contract farming. This focus obscured market-competitive forces which degrade labour conditions and natural resources crucial for local livelihoods.

The key term 'marginal land' likewise concealed crucial uses of land and water by rural populations (Econexus et al, 2009). As a high-profile example, jatropha was originally celebrated as a 'miracle crop' for producing biofuel on 'marginal land', especially in arid areas. In reality, jatropha needed substantial water supplies and even agrochemicals for commercially viable production in Mozambique (Ribeiro and Matavel, 2009). Nevertheless EU biofuel policy presumed a socially beneficent agro-industrial development using 'marginal land'.

Disputing harms from indirect land use change

Throughout the 2008–2009 legislative process of the draft RED, indirect effects through land-use change remained controversial: 'Indirect land use change [ILUC] could potentially release enough greenhouse gas to negate the savings from conventional EU biofuels', warned the Commission's expert body (JRC, 2008: 10). To address ILUC effects, there were proposals to include an extra calculation penalizing all or some feedstocks, thus disqualifying some from the mandatory market for 'sustainable biofuels'.

As a politically awkward issue, any such penalties were deferred. Under the RED, by December 2010 the Commission had to report on ways to calculate ILUC and to minimize its impact (EC, 2009: 40). Further debate over ILUC depoliticized the issues and reinforced the target.

Indirect Land Use Change (ILUC) effects: carbon savings or a carbon time-bomb?

The European Commission's DG Trade invited the International Food Policy and Research Institute (IFPRI) to carry out an ILUC study, especially on potential effects of the EU target and specific feedstocks. It concluded that the 10 per cent target need not undermine GHG savings because conventional (first-generation) biofuel crops would need to provide only 5.6 per cent of transport fuel from renewable energy. This prediction assumed that nearly half the 10 per cent target would come from other renewable sources, especially second-generation biofuel crops and electric cars powered by renewable electricity (IFPRI, 2010: 45).

Yet this prediction was more optimistic than the modest expectations of the motor vehicle industry for electric cars (Harrison, 2010). It also contradicted policy makers' assumptions that 'renewable' transport fuel would mean mainly biofuels by 2020 (cited in Sharman, 2009; cf Bowyer, 2010). By making its assumptions explicit, the IFPRI study revealed weaknesses of beneficent claims for the EU targets to reduce GHG emissions. Its optimistic assumptions provoked disagreements in finalizing the report and again afterwards in publicly interpreting the results, especially in stakeholder meetings held by the Commission (Harrison, 2010).

Environmental NGOs warned that biofuel expansion would trigger great ILUC effects, framed by ominous metaphors such as a carbon-debt time-bomb. Criticizing the IFPRI study, they counterposed pessimistic assumptions about the EU's future dependence on conventional biofuels (for example, T&E, 2010a: 1). Such critics warned that many decades or even centuries may be needed to repay the 'carbon debt', which would be much greater according to less optimistic assumptions. This debt is ignored by 'carbon laundering' under statutory criteria which account only for direct changes in land use (T&E, 2010b). A similar warning deployed financial metaphors: 'The EU is taking out a sub-prime carbon mortgage that it may never be able to pay back' (BirdLife International et al, 2010).

When National Renewable Energy Action Plans were submitted to the European Commission, the aggregate plans contradicted the Commission's optimistic assumptions; some member states expected nearly the entire 10 per cent to come from edible biomass, that is, conventional biofuels. Citing the new evidence, the Institute for European Environmental Policy (IEEP) issued a report warning that the EU's 10 per cent target would generate much greater GHG emissions than indicated in the IFPRI report. In particular, conventional biofuels would contribute up to 92 per cent of total biofuel use, representing 8.8 per cent of the total energy in transport by 2020 – by contrast to only 5.6 per cent in the IFPRI report (Bowyer, 2010).

Moreover, '72% of this demand is anticipated to be met through the use of biodiesel', by contrast to the 55 per cent presumed by the IFPRI report. Biodiesel indirectly increased demand for Asian palm oil, in turn destroying peatland and forests, thus generating greater ILUC effects than bioethanol does (Bowyer, 2010: 2). As the Commission's study had warned, such destruction undermines GHG savings (IFPRI, 2010: 26).

Consequently, biodiesel increases would contradict the environmental rationale for the EU's 10 per cent target (see protest against biodiesel, Figure 4.2). According to the IEEP report, 'The current evidence clearly points to ILUC emissions undermining the arguments for the use of conventional biofuels as an environmentally sustainable, renewable technology'. As a way forward, the report advocated greater consensus on assumptions in modelling ILUC effects (Bowyer, 2010: 21). Campaign NGOs saw the report's expert status as a means for such criticisms to gain greater authority and be taken seriously in the ILUC debate (interviews with environmental NGOs, 10 May 2011 and 17 May 2011).

Citing the IEEP report, nine NGOs jointly criticized the EU's 10 per cent target in a campaign brochure, Driving to Destruction: 'The sustainability of national and European biofuel targets must be reviewed to reflect the reality of biofuel expansion on total emissions, biodiversity and communities' (BirdLife International et al, 2010: 4). Together they demanded broader sustainability criteria, especially by counting ILUC. They cited Southern NGOs' reports on how land-use changes were harming the environment and rural populations.

Figure 4.2: Protest at Biodiesel Expo 2007

Source: Biofuelwatch

Such demands faced several obtacles. First, expert models had an inherent complexity that helped industry to raise doubts about pessimistic scenarios of harmful changes. Second, by contrast with the 'food versus fuel' issue, NGOs foresaw difficulty in explaining the ILUC concept to mobilize supporters for protest. Third, Southern governments were deploying diplomatic means to oppose more stringent sustainability criteria as a threat to their exports (interview, NGO, 17 May 2011). As another awkward reason, some large NGOs were internally divided between climate campaigners originally supporting the draft Directive versus social justice campaigners later opposing it.

For all those reasons, the EU policy framework and consultation procedure marrginalized controversy over agrofuel expansion as global plunder. The ILUC focus depoliticized societal conflicts and reinforced the 10 per cent target. Contrary to NGOs' optimistic expectations for influence, such pressures did little to strengthen the sustainability criteria, much less to weaken the target.

'Investment climate' for bioeconomy innovation

As a further step towards its ILUC report, the Commission held a public consultation, which basically asked: Does the available knowledge justify extra regulatory criteria? Through the consultation procedure, dissent over the EU's optimistic assumptions was further channelled into arguments about the need for better carbon accounting methods. Dominant arguments emphasized the need to maintain financial-regulatory incentives for investment in novel biofuels which would avoid any harmful effects of land-use change.

After its public consultation, the Commission report reiterated the EU's beneficent expectations for the 10 per cent target, linked with promises for technological innovation (CEC, 2010c: 2). Hence any extra regulatory criteria must preserve incentives for investment, according to the Commission. Its report reiterated that the RED creates a 'stable and predictable investment climate', which 'needs to be preserved', especially to stimulate advanced biofuels (CEC, 2010c: 14). Within that aim, the Commission listed three medium-term choices: continue to emphasize deficiencies in modelling, or increase the general requirement for GHG savings, or impose a GHG penalty on some biomass feedstock.

Those ILUC options provoked expert disagreements, serving as a proxy for political conflicts over the 10 per cent target. Indeed, the target was not designed mainly for environmental protection – merely 'an afterthought to security-of-supply concerns', according to an author of the IFPRI study (Rankin, 2011). Soon a broader network of North–South development NGOs attacked EU policy for its 'energy-based target for agrofuels' (EuropAfrica, 2012: 12). Going beyond sustainability criteria, some campaigned against the 10 per cent target (for example, Action Aid, 2012: 11).

The anti-biofuels campaign provoked disagreement among civil society groups. Some large environmental NGOs concentrated all their lobbying efforts on the 'only realistic aim', namely: that the EU mandate should accurately account for carbon emissions through specific ILUC factors. According to such NGOs, gaining ILUC factors in the RED was 'the only game in town', that is, as the only possible way to limit damage from biofuels. 'They warned us that attacking the 10% target will weaken the Directive, which is globally important progress for renewable energy' (personal correspondence, development NGO, 2 July 2012). Formally speaking, environmental NGOs opposed the 10 per cent target (for example, FoEE, 2012). They discussed with development NGOs how to plan a joint opposition campaign, though this did not emerge.

Nearly two years after its indecisive 2010 ILUC report, the Commission proposed to limit 'the contribution made from biofuels and bioliquids produced from food crops' – to 5 per cent of energy use in the transport sector by 2020. The proposal aimed to 'limit the contribution that conventional biofuels (with a risk of ILUC emissions)' make to RED targets, while also 'protecting existing investments until 2020' (CEC, 2012: 2–3). As the latter phrase indicated, the timescale might not deter an increase of biofuel production from food crops before 2020.

A potential limit on biofuel feedstock from food crops provoked criticisms from both sides. NGOs criticized the Commission proposal as insufficient to ensure that the RED targets reduce GHG emissions and avoid various environmental harms (T&E, 2012b). By contrast, according to biofuel advocates, a 5 per cent limit would be unfeasible and jeopardize prospects for a European bioeconomy. 'Advanced biofuels need to be developed and subsequently deployed at the same time as continuing to develop sustainable conventional biofuels beyond 2020'. Together they would be essential for 'developing the bioeconomy, which provides a new "green growth" opportunity for European farmers, foresters, fishermen and their cooperatives', argued the agri-industrial farm lobby. It opposed any ILUC restrictions on feedstock (COPA-COGECA, 2013b: 2).

North–South development NGOs sought to popularize their attack on the EU biofuel mandate. A satirical film, *Drive Aid*, asked people whether Southern rural populations should sacrifice their natural resources so that European motorists can avoid fossil fuels (Action Aid UK, 2013). Development NGOs targeted the EU's KBBE framework and its industry lobby groups, which together were driving the EU's biofuel mandate for global resource plunder (Paul, 2013: 28–29). Given that agenda,

> There is no point in having a complex agroenergy policy framework with targets, subsidies and other incentives, plus environmental and social criteria, along with an ILUC factor, if the fundamental truth about agroenergy is that it is not renewable and never will be. The whole agroenergy policy framework should therefore be dismantled. (Paul, 2013: 31)

To benefit Southern rural populations, peasants moreover, must help to shape agricultural policies based on sustainability, food sovereignty and agroecology (Paul, 2013: 25). Local popular control over agri-food systems illustrates wider eco-localization agendas.

Some critics sought to mobilize other NGOs against the EU's bioeconomy agenda:

> Bioeconomy policy documents highlight the need to accommodate the ever-increasing call for bio-products and biomass, rather than suggesting alternatives that could decrease demand. This means that more and more land will be converted to multiple-use 'flex crops' like soy, sugar and corn, often at the expense of other food crops. This trend – creating new biologically 'enhanced' products as well as new ways for humans to take control over resource production – leads to the commodification of nature. Furthermore, it perpetuates structures

that prioritise market growth over environmental health and human wellbeing. (Hall and Zacune, 2012; cited in TNI, 2015: 5–6)

This attack on a key driver remained marginal among civil society groups.

Bioeconomy imperative supersedes indirect land use change (ILUC) harms

GHG reductions were less important than other aims for some supporters of the EU mandate. These priorities became more explicit to avert stringent conditions on ILUC effects. The industry lobby asked the EU 'to make biofuel policy effectively contribute to the security of energy supply alongside the future European bioeconomy'. The EU should recognize that most biofuels available by 2020 would still be conventional ones, it argued (EBTP, 2015). Industry's arguments normalized harmful ILUC effects from oilseed crops as an acceptable baseline.

After several years of such debate on tightening the RED criteria, the 2018 revision increased the mandate to 14 per cent renewable energy in transport fuel by 2030. Advanced biofuels had to comprise 0.2 per cent of transport fuel in 2022 and gradually increase thereafter (EC, 2018). Yet they still lacked commercial viability a decade after the 2009 RED was meant to incentivize their development. Cellulosic biomass had been promised as the basis for long-term environmental sustainability; this promised served to justify enormous EU budgets for their technoscientific development, alongside expansion of conventional biofuels (Ernsting and Smolker, 2018).

As the most contentious feedstock, the EU's palm oil imports were rising, especially from Asian countries which were notorious for plantations destroying forests. Half of all these imports became biodiesel feedstock as a relatively cheap, high-energy means to fulfil the EU mandate. Palm oil had been named as 'high ILUC-risk' feedstock in a draft of the 2018 revised Directive. Nevertheless its risk status was left ambiguous in the final version, thus protecting palm oil imports and their commercial advantage over the EU's domestic oilseeds. According to a trade expert, some Commission officials were willing to 'throw EU farmers under the bus in their anxiety to do grubby trade deals' (Michalopolous, 2018).

For the 2018 Directive, negotiations reached an ambiguous compromise over 'high ILUC-risk' crops. Such feedstock 'shall not exceed the level of consumption of such fuels in that Member State in 2019' through the year 2023 and must decline thereafter to zero by 2030 (EC, 2018: 126). However, lax criteria at national level could allow palm oil to double its role in EU biofuels, facilitated by trade negotiations with Asian countries (Morgan, 2018).

Palm and soy oil meanwhile remained prevalent in biofuel feedstock, thus undermining the putative environmental benefits. Biofuels from those feedstocks caused three times more carbon emissions than would have been emitted if vehicles instead used fossil diesel (T&E, 2021). When taking into account carbon impacts of ILUC, the EU's average GHG intensity of transport fuels in 2018 was only 2 per cent lower than in 2010 (EEA, 2020). This scant improvement arose from political processes greenwashing the EU's resource plunder and GHG emissions, while leaving no one responsible for the various harms.

Conclusion

Let us return to the questions presented at the start of this chapter:

- How did the EU justify and create a new market for biofuels? Through what institutional changes and beneficent societal vision?
- How did the EU biofuels mandate provoke opposition, controversy and alternative agendas? With what outcomes?
- How do these divergent agendas relate to system change versus continuity?

Emphasizing systemic forces, those questions extend an investigation that was originally done through Participatory Action Research (Franco et al, 2010; Levidow and Oreszczyn, 2012).

The EU mandate for 'sustainable biofuels' has been analysed here as a techno-market framework. It had several institutional commitments: to protect farmers' income, to make the EU a buyers' market for competing biomass suppliers, to finance R&D for future biofuels which could gain intellectual property, to strengthen the 'investment climate' for techno-innovations, to promote a KBBE, and to normalize the rising transport of the EU's internal market. Such political-economic aims expressed an economic imaginary of a globally more competitive Europe; all this took priority over any climate benefits.

According to promoters of the EU mandate, biofuels would substitute for oil, reduce GHG emissions, strengthen energy security and enhance rural development. Early on, those beneficent claims provoked a public debate, initially focusing on 'food versus fuel' conflicts over land use. In response, a state–industry alliance anticipated 'second-generation' (or advanced) biofuels, whose feedstock would use non-food biomass, grown on 'marginal land'. These promises justified state funds for companies to develop such low-carbon eco-efficient innovation.

From an STS co-production perspective, the EU agenda co-produced distinctive forms of nature (decomposable and recomposable biomass; marginal land as abundantly available), technoscience (an 'investment climate' for corporate R&D capitalizing bioresources) and social order (beneficent market competition via rising private transport and technoscientific R&D).

An EU–industry alliance co-produced those complementary forms in promoting the KBBE.

This sociotechnical imaginary facilitated the EU's rising GHG emissions, namely: The initial market for conventional biofuels was a prerequisite for market incentives to stimulate advanced ones. As an alibi for fuel inefficiency, the motor industry lobby and its state supporters cited future GHG savings from biofuels. In practice the EU biofuels mandate incentivized efforts to appropriate fertile land and scarce water, often by dispossessing rural communities in the global South. The techno-market framework avoided or blurred political responsibility for such outcomes.

For all those reasons, the EU's mandate agenda was opposed firstly in 2007 by North–South civil society networks which take up rural development and socio-environmental justice issues. They denounced 'agrofuels' for land-use changes degrading natural resources, competing with food needs and undermining livelihoods in the global South. This critical perspective served as a bridging frame, aligning diverse frames and emphases of civil society groups.

More civil society groups soon criticized the biofuel mandate in the draft RED as a false solution, especially because it depended on a false method of carbon accounting. Although large environmental NGOs had originally supported the draft RED as a climate measure, they were soon persuaded to criticize the EU mandate for inadequate or deceptive sustainability criteria. The EU's biofuels mandate was seen as a multiple threat by various stakeholder groups and so became more vulnerable to political attack. By 2008 relevant large NGOs opposed the EU mandate within the draft RED before it was enacted.

Responding to such criticisms, the EU–industry biofuel alliance framed any potential harm as manageable through sustainability criteria. The 2009 RED specified biomass eligibility for EU targets – for example, what environments can be further industrialized for eligible biofuels, what must be the minimum GHG savings, and which biofuels can gain a GHG-savings bonus. Carbon accounting selectively quantified some environmental harms, while disregarding others. The criteria decontextualized land-use changes from any local population and its resource usages.

Beyond direct changes in land use for biofuel crops, the EU's mandate was stimulating ILUC, especially for oilseed production, which greatly increased GHG emissions and environmental degradation in the global South. According to several large environmental NGOs, a false carbon accounting concealed the enormous 'carbon debt' and consequent 'carbon time-bomb' from biofuels; hence the EU mandate must adopt ILUC criteria to exclude or restrict the most harmful feedstock, they argued. Rather than do so, EU procedures channeled such issues into expert consultations about ILUC carbon-accounting models, which became imore contentious and complex. Alongside opposing the biofuels mandate, some NGOs pursued sustainability criteria that would limit or exclude the most harmful feedstock.

These efforts faced several obstacles. As the most important, any ILUC restrictions on biofuel feedstocks were opposed by lobby groups for relevant industries and agri-industrial farmers (for example, COPA-COGECA, 2013b). Some environmentalist politicians anticipated that opposing or limiting the biofuels mandate could jeopardize the wider mandate for renewable energy in the overall 2009 RED. Large environmental NGOs had internal tensions between agri-development campaigners versus climate campaigners, who anticipated technological means making biofuels environmentally sustainable. The various critical groups had diverse alternative proposals, which could not be readily aligned (Table 4.1; cf Snow et al, 1986). Relative to the 'food versus fuel' dispute, the ILUC issues seemed more complex for explaining to supporters.

Without an active mobilization of their Europe-wide supporters, agrofuel critics had difficulty to gain more stringent sustainability criteria, much less to undermine the biofuel mandate. The EU's resource plunder was depoliticized through technical criteria for 'sustainable biofuels', in turn channelled into expert disagreements over ILUC factors. Some groups eventually criticized the EU's bioeconomy agenda as a key driver of agrofuels; some counterposed a public-goods form of bioeconomy (Schmid et al, 2012; Paul, 2013; TNI, 2015). Beyond biofuels, the RED's incentives for large-scale bioenergy were opposed by a North–South network of civil society groups, but minus the large European environmental NGOs (Bioenergy Out Declaration, 2015).

More than a decade after the 2009 RED promised long-term sustainability through 'advanced biofuels', such an innovation remained elusive. Yet the promise still justified the EU's commitment to biofuels and large R&D expenditures benefiting high-carbon industrial sectors (Ernsting and Smolker, 2018). Meanwhile the EU's techno-market fix made the state more dependent on anonymous markets and private-sector actors, thus blurring political responsibility to fulfil climate targets and ensure environmental sustainability. This arrangement strengthened dominant economic actors in the fuel and motorcar sectors, aiming to protect an 'investment climate' for bioeconomy innovation; this favoured system continuity rather than protecting the climate.

It was difficult for opponents to undermine the biofuel mandate without opposing its sociotechnical imaginary of a future Europe. Despite the difficulty, critics highlighted how the agri-industrial system threatens land tenure, rural livelihoods, carbon sequestration and natural resources, especially in the global South. Many civil society groups linked all these issues through an alternative agenda for food sovereignty through agroecology. This offers a potential contribution to system change.

UK Waste Incineration Fix: Perpetuating and Displacing Waste Burdens

Introduction

For several decades there have been global conflicts over waste disposal, often a euphemism for dumping or burning waste. In late 20th-century Europe, a default mode was landfill, whose methane emissions are a potent GHG. Towards improvements, there have been policy agendas to reduce or reuse waste rather than dispose it.

The 2008 EC Waste Framework Directive formalized the waste hierarchy. This ranked management steps from best to worst – prevent, prepare for re-reuse, recycle or other recovery, and lastly disposal. The Directive brought 'a modernised approach to waste management, marking a shift away from thinking about waste as an unwanted burden to seeing it as a valued resource' (EC, 2008, 2010: 5). The framework sought innovative solutions for decoupling 'the link between economic growth and the environmental impacts associated with the generation of waste' (EC, 2008: paragraph 40; EC, 2010; cf EEA, 2009).

Through the waste hierarchy, the EU framework integrated the 'alternatives of reducing waste and extracting value from it' (Corvellec and Hultman, 2011: 5–6). As an instrument of ecological modernization, the framework stimulated a search for technoscientific innovation which could reconcile economic growth with lower resource burdens. The EU's policy framework expected waste authorities to manage private-sector competition for contracts, thus potentially blurring responsibility to interpret and implement the waste hierarchy. 'Municipal waste-management companies perform the fix' (Hultman and Corvellec, 2012: 2418).

Throughout the EU, waste-management systems have undergone pressures to move waste beyond mere disposal, towards valuing waste as a resource.

Some governments expanded mass-burn incinerators, claiming that the electricity generation was resource recovery. Such programmes provoked protests and public controversy.

According to such opponents, incineration programmes perpetuate a system that wastes resources and generates GHG emissions. As an alternative, a circular economy would redesign production systems to minimize and reuse waste (EMF, 2013). A circular economy has been espoused by the European Union and many member states, generally as a technical-managerial agenda relegating responsibility to companies (EC, 2014, 2015).

Within that EU framework, the UK has had a long-time controversy over waste conversion. Relevant UK policies can be understood as a techno-market framework. This creates market competition to generate techno-innovations that could reconcile the diverse objectives of waste management, especially through waste conversion. These techno-optimistic assumptions have served to perpetuate systems that waste resources and displace responsibility.

This study had frequent exchanges with anti-incineration activists, shaping research questions about false solutions. This chapter takes up the following questions:

- How were market-type incentives meant to stimulate novel technologies fulfilling environmental policy aims, especially bringing waste up the hierarchy and saving GHG emissions?
- How was techno-innovation meant to go beyond incineration and overcome its limitations?
- How did protest dispute those claims and use the opportunity for alternative agendas?
- How does this conflict relate to system change versus continuity?

The chapter has the following sections: the UK Energy-from-Waste (EfW) policy framework as a techno-market fix, generating public controversy; ATTs, claiming to go beyond incineration yet extending the controversy; incineration becoming more contested around resource and climate issues; and, finally, a conclusion answering these questions.

UK disputes over waste incineration versus alternatives

Within the EU policy framework to reduce waste landfill, the UK subsidized EfW plants, mainly a euphemism for incineration, the most controversial type. Underlying their expansion was an ecomodernist policy framework. This section describes the initial UK controversy.

UK ecomodernist framework for waste-as-resource

For its low-carbon strategy, the New Labour government (1997–2010) sought to reconcile environmental aims with economic interests. To deter waste disposal, the Landfill Allowance Trading Scheme imposed penalties for any local authority exceeding its maximum allowance for landfilling biodegradable waste; any surplus allowances could be traded with other local authorities, thus creating a market. The UK's Landfill Tax Escalator set a timetable for annual tax increases, rising sharply since 2005 and quadrupling in the subsequent decade; this rise drove up the gate fees paid to waste-management companies. All this incentivized waste diversion to incinerators, which had no analogous tax.

The policy framework attributed the waste problem to 'market failure', warranting support measures for better market competition that would stimulate eco-efficient innovation (HMG, 2009: 136). The state's role was put more bluntly in the commonplace mantra as regards technology choices: 'The government allows the market to decide' (quoted in Levidow and Papaioannou, 2016). This techno-market imaginary blurred responsibility for fulfilling environmental policy aims.

UK policy has promoted the waste hierarchy as a 'guiding principle' for new facilities (DEFRA, 2007a: 2; Figure 5.1). Along ecomodernist lines, an investment metaphor linked the environment positively with economy: 'The dividends of applying the waste hierarchy will not just be environmental. We can save money by making products with fewer natural resources, and we can reduce the costs of waste treatment and disposal' (DEFRA, 2007b: 9).

Alongside renewable energy, the UK government sought to reduce GHG emissions from waste-management practices through joint solutions with industry. Implementing the 1999 EC Landfill Directive, the UK's Landfill Tax Escalator set a timetable for annual tax increases that rose sharply starting in 2005 and quadrupling in the subsequent decade; this drove up the gate fees paid to waste-management companies. This rise stimulated efforts towards waste reduction and kerbside segregation, which in turn facilitated recycling and organics segregation; the latter provides inputs for a composting process or anaerobic digestion plants.

There were many efforts at source-segregating waste categories, thus minimizing heterogeneous MSW. Nevertheless, large quantities still needed an outlet. For this residual MSW, UK policy promoted EfW plants: 'Generating renewable energy from biomass waste could also significantly reduce the amount of waste that is landfilled in the UK' (HMG, 2009: 87, 104).

To help local authorities to fulfil their statutory duty for landfill diversion, from 2006 onwards the UK's Private Finance Initiative (PFI) scheme offered subsidy for a Waste Infrastructure Delivery Programme. The policy explicitly favoured recyclables removal through long-term contracts with

Figure 5.1: Waste hierarchy

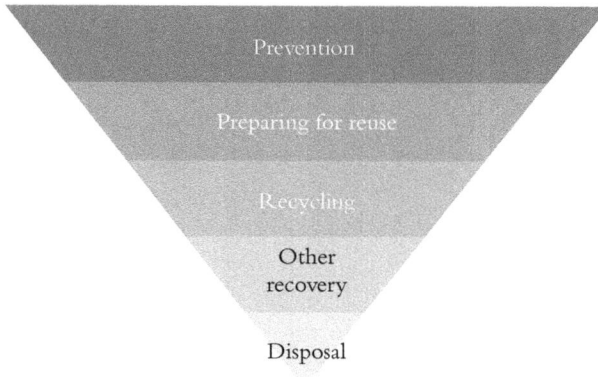

Source: DEFRA (2011a)

waste-management companies. Yet the PFI scheme mainly subsidized contracts for large mass-burn incinerators to treat residual MSW, whose throughput ranged between 60 and 300 kt/year (see Figure 5.2).

For its environmental aims, UK policies have relied on the 'right' price signals for business, and local authorities outsourcing waste management. Consequently, new commercial markets turned waste into a resource mainly for large incineration plants. Waste-treatment systems shifted to a larger scale, remaining distant from any specific end-use. According to its advocates, the public good is globally served by turning waste into a useful resource somewhere, anywhere. This meant mainly converting waste into electricity (Reno, 2011). Yet such claims for global goods became contentious as regards GHG reductions and the waste hierarchy. Some uses were rivalrous; electricity incentives often undermined environmentally preferable uses of the same resource (Alexander and Reno, 2014: 351–354).

These tensions arose from the policy framework. 'Government policy is driven by the desire to drive waste up the hierarchy' (DEFRA, 2014a: 67). As the underlying problem-diagnosis, environmental externalities such as landfill emissions result from market failures, warranting policy intervention to adjust market competition (DEFRA, 2011b: 4, 8). Such policy language has blurred responsibility for outcomes.

Linear versus circular economy

Through the PFI programme, more waste was diverted to new incinerators, which often faced public opposition. Even beforehand, environmental NGOs had opposed incinerators for wasting resources and increasing GHG

Figure 5.2: Mass–burn incineration process

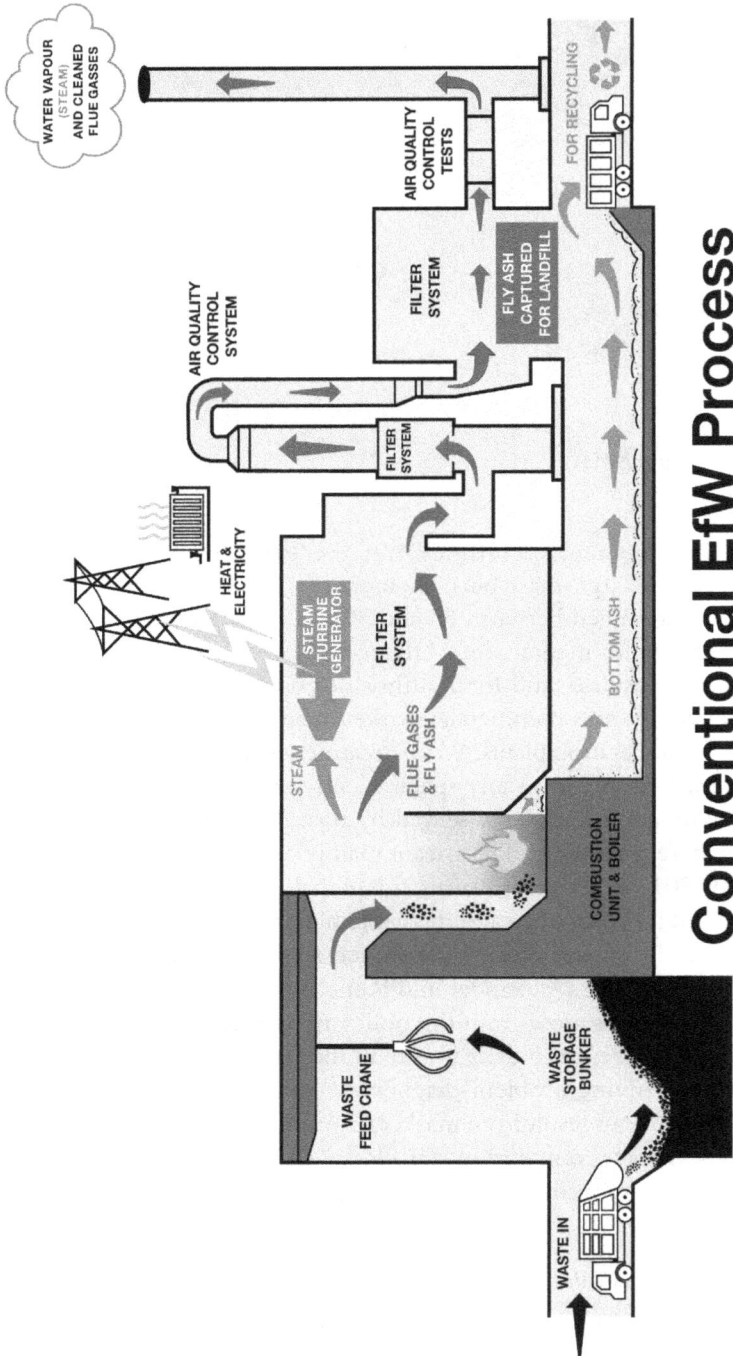

WATER VAPOUR (STEAM) AND CLEANED FLUE GASSES

AIR QUALITY CONTROL TESTS

FILTER SYSTEM

FLY ASH CAPTURED FOR LANDFILL

FOR RECYCLING

AIR QUALITY CONTROL SYSTEM

FILTER SYSTEM

HEAT & ELECTRICITY

STEAM TURBINE GENERATOR

FILTER SYSTEM

STEAM

FLUE GASES & FLY ASH

BOTTOM ASH

COMBUSTION UNIT & BOILER

WASTE FEED CRANE

WASTE STORAGE BUNKER

WASTE IN

Conventional EfW Process

Source: www.arc21.org.uk/

emissions (Let's Recycle, 2004). They counterposed a segregation process to optimize resource recovery (FoE, 2006: 5; citing Eunomia, 2006). Some protests demanded alternative solutions to minimize waste transport (Dodds and Hopwood, 2006; Rootes, 2009). Although these protests rarely closed down plants, the conflict deterred some local authorities from commissioning new incinerators and pushed them towards alternatives with greater local responsibility.

In the period when the UK's techno-market policy framework was stimulating new investment in waste management technologies, especially incinerators, an expert–environmentalist alliance instead was promoting a circular economy. This would require 'ever-decreasing circles' of materials usage, especially through disincentives against incineration and against Mechanical and Biological Treatment designs which generate feedstock for incineration. It proposed taxing such plants, 'with a view to further incentivising movement of waste up the hierarchy'. Beyond market adjustments, the alliance also advocated producers' legal responsibility for materials recyclability and green criteria for public procurement (RWM, 2014: 9; 19; see also CIWM, 2014).

A Parliamentary Committee likewise advocated a circular economy through system redesign: 'Reducing the dependency on primary resource use for economic growth is an essential part of moving to a more sustainable economic system.' This would be incentivized by 'eco-design standards across a range of products', including a ban on non-recyclable materials (EAC, 2014: 33; citing EMF, 2013).

Environmental NGOs advocated greater waste reduction and recycling, while stigmatizing incineration as a 'waste of resources'. Although often sharing this perspective, local opposition groups strategically focused on issues considered by local and planning authorities – especially air pollution, health risks and siting (Rootes, 2009). Local campaigns sometimes persuaded their local authority about health hazards, but this issue was officially the preserve of the Environment Agency, which generally accepted safety claims. As another difficulty, siting issues could pit local campaigns against each other. As a way forward, one campaign successfully demanded greater recycling in order to block a new incinerator (Dodds and Hopwood, 2006).

The national anti-incineration network diagnosed the systemic problem as the linear economy – 'make anew, use, dispose' – which perpetually generates waste. It counterposed a circular economy, which would restructure production processes to reduce waste generation at source (UKWIN, 2016b).

The traditional 'linear economy' is one based on extraction and processing, followed by consumption and disposal. In contrast, in the circular economy materials are neither burned nor buried, products are designed to be re-used and recycled and repaired, and nutrients are retained. One needs no 'end of pipe' technology such as incineration, because the pipe never ends. (UKWIN, 2015: 2; also UKWIN, 2016b)

This contrast between 'circular versus linear economy' served as a bridging frame. It aligned diverse issues from local campaigns, while going beyond expert disputes over biophysical effects. However, UK policy incentives still stimulated more thermal treatments for MSW while marginalizing alternatives, thus provoking national and local conflict.

Gloucestershire conflict: rival designs for resource recovery

As a national network, UK Without Incineration Network (UKWIN) built on numerous local campaigns opposing incinerators and often counterposing alternative designs. Such campaigns have linked issues of resource recovery, democratic accountability and societal benefit. Those widespread dynamics are illustrated here by a local dispute between two political forces with rival designs for resource recovery.

In 2012 the Conservative-controlled Gloucestershire County Council signed a contract with Urbaser Balfour Beatty (UBB) for an incinerator at the Javelin Park site, despite strong opposition from community groups and other political parties. A Labour councillor stated, 'It is immoral that the County own the land, they're letting the contract and it looks like they are going to give themselves planning permission. We are supposed to live in a democracy' (BBC, 2012).

After the Conservative Party lost its majority in the 2013 local election, the new Council refused planning permission for the UBB plant and started to devise an alternative. But UBB appealed the reversal and was supported by the government (Reece, 2013; Merrell, 2019). The Council had a legal difficulty to break the contract, so the plant construction went ahead.

Opposition was led by Gloucestershire Vale Against Incineration, which organized objections to the UBB's planning application in 2015. Small local campaigns united to support an alternative, an R4C plant – Resource Recovery, Refining and Recycling Centre – based on Mechanical, Biological and Heat Treatment. According to its designer, 'This is a people power project. We deserve to have something we want and I hope we can use the energy shown by campaigners for something positive', thus giving 'energy' a social meaning. By contrast, the campaign denounced the UBB design as an expensive 'dinosaur technology' (Perchard, 2015).

Community R4C promoted the alternative design as an 'advanced' plant, with a lower carbon footprint than the UBB plant:

> CommunityR4C is a community-led initiative to provide economic, social and environmental benefits for Gloucestershire by treating our

waste as a valuable resource rather than burning or burying it. Our main aim is to help build an advanced, commercially sustainable recovery and recycling plant which can turn more than 90% of the county's waste into valuable materials for sale and re-use.

This design was presented at public meetings and depicted in a short film, presented by the actor Jeremy Irons (Community R4C, 2015). The campaign soon added its support for a circular economy in Gloucestershire.

When the UBB incinerator became operational, the company appropriated the language of its opponents: 'The Gloucestershire Energy from Waste Facility is proud to champion the 4Rs: Reduce, Reuse, Recycle and Recover.' A Conservative Party councillor praised the plant as 'a cost-effective and environmentally sound solution for processing the county's waste that can't be recycled', reinforcing fatalistic assumptions (UBB, 2020). Thus rival agendas gave divergent meanings to the waste hierarchy, especially 'resource recovery'.

The local MP for Stroud, David Drew, had been elected in 2017 on an anti-incineration platform. He supported the campaign's demand for a public inquiry on the Council's contract for the plant, whose cost was rising. Like many other MPs, he raised such issues in the House of Commons (see penultimate section, Hansard, 2020; UKWIN, 2020).

By 2019 the campaign was joined by XR activists. Protesters blocked traffic on the access road to the EfW facility and burned fake money, symbolizing the funds being wasted (Merrell, 2019). According to protestors, the incinerator was damaging to health, the environment and Council funds. Over half the waste being burned could be instead recycled with an alternative plant. When obtaining confirmation that the UBB plant had no pre-sorting of recyclable waste, the campaign issued its own 'Stop Notice', while demanding that the Environment Agency enforce it.

The campaign also organized crowd-funding for a judicial review of the Council–UBB contract (GL Law, 2019). According to the High Court, the Council had hidden the true cost of its largest contract – not only from taxpayers, but also from its own councillors. The campaign also sought disclosure of the auditor's report, so that the public could judge whether the UBB plant was 'value for money'. But the Conservative-led Council conveniently delayed publication until after the 2021 local election.

This local conflict illustrates some nation-wide patterns: Both sides cited the waste hierarchy and 'advanced' technology to justify their contrary designs. Protest gave greater prominence to an alternative design, which helped to stigmatize incineration for wasting resources rather than recovering them. Anti-incineration campaigners framed the issue as democratic accountability of a plant design to benefit the community rather than feed electricity to the national grid. The two designs had sociotechnical features serving divergent interests and societal visions. Hence this dispute exemplifies rival sociotechnical imaginaries of a future society (see again Table 5.1).

Technological advance beyond incineration?

The UK government supported industry efforts to bring MSW treatment beyond conventional incinerators and thus disposal. Support measures stimulated private-sector investment in ATTs, which provoked suspicion and rarely delivered improvements. Controversy over incineration was extended rather than overcome.

For ATTs, a priority was a gasification process to treat heterogeneous MSW. Some designs were classified as 'advanced', producing a synthetic gas (syngas). As the most ambitious aim, a gasifier would clean the contaminants, so that the syngas could substitute for fossil fuels in transport. Advanced technologies 'have the potential to deliver more efficient generation in the long term and have the potential to deliver further benefits beyond renewable electricity generation', especially through a clean syngas that can substitute for fossil fuel or be used as a chemical feedstock (DECC, 2012: 72), thus bringing waste beyond mere disposal (DEFRA, 2013b, 2014a).

With such techno-optimistic expectations, government support 'means that the UK has become an internationally appealing market for the development of energy-from-waste projects using newer gasification and/ or pyrolysis technologies' (GIB, 2014; see also ETI, 2012). Through circular reasoning, state investment in sub-optimal technology was necessary to gain inward investment for better technology.

But some experts raised caveats about difficulties. In particular, 'gasification of wastes continues to face several technical and economic issues, mainly related to the highly heterogeneous nature of feeds such as municipal solid wastes' (Arena, 2011: 406; also Rollinson, 2019). To develop gasifiers, the innovation has depended on support measures at several stages – for example, R&D funds, demonstration plants and operational subsidy.

According to their proponents, gasifiers improve the negative features of incineration – by keeping waste-flows local and minimizing bottom-ash disposal – while also matching the positive features, especially its reliable operation and financial bankability. Such optimistic visions for gasifiers helped to mobilize resources, especially policy support and R&D investment. Moreover, 'advanced' plasma-gasifiers promised greater benefits, helping to justify state support. Whenever encountering technical obstacles to those benefits, optimistic expectations have been readily shifted to future technologies 'beyond incineration'.

Proponents emphasized ATTs' flexibility. This 'enables production of renewable heat and power, fuels, gases such as hydrogen, and/or chemical intermediates', according to the Renewable Energy Association (REA, 2014).

Towards such ambitious aims, some technology development companies designed high-temperature plasmafication techniques to clean the syngas, drawing analogies with hydrogen fuel cells (Air Products, 2011; APP, 2013), thus going beyond incineration. Yet the designs generally resulted in technical failure and great financial loss, most notably a £1 billion loss for the Tees Valley gasification plants, which never functioned properly (Air Products, 2016). As a result, Air Products wrote off up to £1 billion (Tighe, 2016). As a recurrent pattern, experimental plants had great difficulties in treating variable heterogeneous feedstock, often malfunctioned and underwent great delays.

In the public controversy, each side framed gasification differently vis-à-vis incinerators. Advocates favourably represented them as beyond incineration, thanks to design improvements. Opponents pejoratively represented all thermal treatments as incinerators, perpetuating their negative features such as burning waste, wasting resources and undermining efforts to minimize waste production (UKWIN, 2010). Expert submissions dissuaded the Environment Agency from issuing environmental permits for some gasifiers, for example, at Castleford and Cwmgwili in 2014 (UKWIN, 2016a).

Thus controversy was extended to the technological design, evaluation criteria and expert judgements. In such ways, claims to advance beyond incineration have been contentious as regards three main issues: health and amenities issues (DEFRA, 2013b: 32), the waste hierarchy and waste recovery. The rest of this section focuses on the latter two issues, which also matter for GHG emissions.

Advanced Thermal Treatments moving waste up the hierarchy?

Incineration has increasingly faced criticism for wasting resources and so contradicting the waste hierarchy. The latter has been officially 'a guiding principle' for new treatment facilities (DEFRA, 2007a: 2). 'Government policy is driven by the desire to drive waste up the hierarchy', for example, through EfW plants (DEFRA, 2014a: 67). As the government eventually acknowledged, incineration may conflict with recycling: 'lower rates could result from an authority focusing on avoiding landfill by investing in incineration and targeting its waste management policies on that treatment solution, rather than poor recycling awareness or initiatives' (DEFRA, 2012: 4).

Bringing waste up the hierarchy would depend on 'suitably flexible facilities and contracts – that is, that do not "lock in" an unreasonably high proportion of waste, should waste prevention, reuse and recycling performance substantially increase' (DEFRA, 2013a: 5).

At the more local level, the risk that energy from waste can compete with, not complement, recycling does exist. However, it is an avoidable risk if contracts, plants and processes are flexible enough to adapt to changes in waste arisings and composition. (DEFRA, 2014a: 3)

The focus on flexibility evaded the issue of incentives for feeding thermal plants with high-calorific value materials such as plastics.

Indeed, long-term contracts for large incinerators undermine the waste hierarchy, argued campaign groups and some experts. Early on, the PFI programme and its specific plants came under attack from environmental NGOs: 'There is growing concern that the PFI process encourages local authorities to procure large, inflexible facilities such as incinerators, rather than implementing schemes to maximise recycling and provide small-scale, flexible technologies to deal with the waste left after recycling and composting' (FoE, 2008: 1).

According to many critics, incinerators generate commercial pressures to 'feed the beast'. In other words, they perpetuate a long-distance, large-volume feedstock supply to ensure that the plant recoups its investment and remains financially viable. New plants lock in long-term demand for feedstock and so deter greater recycling, argued local campaign groups (Connett, 2013; Seltenrich, 2013; Perchard, 2015). To avoid such outcomes, local waste authorities could consider 'the proximity principle', that is, arrangements localizing waste management (DEFRA, 2014a: 24, 43). Proximity has been a criterion for whether local authorities support a new incineration facility, for example, through planning permission or subsidy.

Linking local campaign groups, UKWIN has opposed all thermal-treatment plants for contradicting the waste hierarchy. It has been affiliated with the Global Alliance for Incinerator Alternatives, arguing that gasification is 'incineration in disguise' (GAIA, 2006). 'People should focus on the exit strategy for incineration, not whether one form of incineration should be preferred over another' (UKWIN, 2010).

Portraying all thermal treatments as incineration, the network opposed them all for wasting financial and material resources:

Incineration depresses recycling, destroys valuable resources, releases greenhouse gases, and is a waste of money. Incineration has no place in the zero waste closed-loop circular economy we should be working towards. (UKWIN, 2010)

An unintended consequence of a ban or restriction just on landfill is further long-term 'lock-in' of compostable/recyclable/preventable material into incineration, which not only runs contrary to the Waste Hierarchy but also represents a loss of valuable resources. (UKWIN, 2014: 1)

Figure 5.3: 'Incineration is part of a linear economy'

Note: The text reads: Extract>Manufacture>Distribute>Consume>Discard>Then the items that were burnt and buried are replaced and the process starts again. Humans currently use finite resources as though they were renewable.

Source: Frances Howe in UKWIN (2016b) from *Everything Goes Somewhere*

Regardless of the specific technology, '[i]ncineration is part of the linear economy', thus impeding a circular economy. It also 'has significantly higher carbon intensity than burning gas or coal', argued UKWIN (2016b; see Figure 5.3).

Industry rejected the criticism by blurring the distinction between incinerators and other waste-conversion techniques, as well as by exaggerating the recycling rates of some countries: 'EfW does not act as a disincentive to materials recovery and recycling. Evidence from Europe indicates that high recycling (including composting) rates can be sustained alongside high energy recovery rates', according to the Renewable Energy Association, whose members include some incineration companies (REA, 2011). It advocated state support for ATTs such as gasification plants to treat MSW.

The industry claim about complementary roles had been contradicted by Denmark incinerating 80 per cent of household waste, thus wasting materials and resources which could have been recycled. Acknowledging its mistake, the government announced a 'Denmark Without Waste' policy, which would reduce incineration and recycle half of all household waste (Danish Government, 2013). This policy reversal was publicized by UKWIN.

Waste localization became an extra argument for MSW gasifiers, especially by Energos (MPS, 2007). Gasifiers were initially small-scale plants, anyway necessary to obtain external finance and investment decisions for a novel technology before it was widely 'proven'. According to an expert agency, gasifiers are commercially scalable: they 'can operate at higher efficiency on a smaller scale than traditional incineration plants' (Spice, 2013).

A similar localization perspective came from the Energy Technologies Institute: 'Most UK communities don't produce enough MSW to be economically viable for current-scale technologies, e.g. incineration. A town scale plant is a major development opportunity [offering] benefits in efficiency and reductions in transport impacts including costs' (Evans,

2014). This became an argument to incentivize small-scale ATTs, especially gasifiers, to treat MSW. Such plants would avoid large-scale demands for waste, while more readily finding nearby users for the heat. Yet small-scale commercial plants remained hypothetical.

Meanwhile the waste-management industry had overt disagreements about the UK policy framework. State financial incentives for ATTs were criticized not only by civil society groups, but also by market competitors. Incineration promoters emphasized the benefits of its associated energy recovery, which gains no inherent advantage from ATTs (CIWM, 2013). According to one company, its 'mass burn' incineration technology is already an ATT, on grounds that its novel low-oxygen combustion process reliably increases the potential for efficient recovery of energy and materials (Sigg, 2014). On similar grounds, a company manager questioned support measures favouring ATTs over conventional incineration: 'ATT is driven by the UK subsidy regime, which perversely gives more support to unproven technologies in the UK residual waste treatment market' (Allin, 2015).

As the prevalent form of ATT, gasifiers rarely delivered their putative benefits, for various reasons. As the UK's market leader for two-stage gasification, Energos built four plants simultaneously, had serious difficulties, encountered cash-flow problems, had disputes with its contractors and failed to obtain completion certificates for staged works. The company went into administration (Energos, 2016).

Such technical difficulties often arose at the construction stage, most notably after Air Products obtained finance for two expensive plants in the Tees Valley (Air Products, 2016; UKWIN, 2016a). As one reason for failure, the technology was scaled up too quickly, thus aggravating the inherent difficulties of treating heterogeneous variable waste (Peake, 2016; Rollinson, 2019). Such failures provoked overt intra-industry disagreements over rival technological designs; these reciprocal criticisms were highlighted by the anti-incineration campaign (UKWIN, 2016a). Rather than make its own factual claims, the 'campaign could exploit a battle between King Kong and Godzilla', as critics metaphorically saw advocates of different thermal treatments.

As another contentious issue, plant designers and operators prioritized electricity production in order to maximize income. Regardless of the specific heat-treatment technology, energy conversion and capture efficiencies are poor if there is no economic use for the heat produced, such as a nearby district heating system. Waste heat was being used in only 2 per cent of the UK's EfW schemes (DEFRA, 2014b), partly because state subsidy, market incentives and distribution infrastructure were much weaker for heat use than for electricity.

As the policy context, subsidy was available from Renewable Obligation Certificates, meant to incentivize GHG reductions, which depended on various optimistic assumptions. Likewise driven by electricity subsidies,

ATT plants marginalized heat uses in practice despite promises to provide them. For waste feedstock not fulfilling the definition of dedicated biomass, the biomass fraction may accrue electricity subsidy through Renewable Obligation Certificates if the waste is processed by anaerobic digestion, gasification or pyrolysis (together known as ATTs); or if the waste is used alongside other fuels and the overall biomass content of the fuel mix is greater than or equal to 90 per cent; or if the plant can provide Combined Heat and Power (DECC, 2014; Ofgem, 2015). Many plants have been called 'CHP-ready', yet the phrase has meant simply a potential, heat delivery rarely becoming a reality.

Recovering resources beyond disposal?

The waste hierarchy distinguishes between waste disposal versus recovery; the latter has greater potential for GHG savings. Under EU criteria, incineration encompasses 'thermal treatment processes such as pyrolysis, gasification or plasma processes insofar as the substances resulting from the treatment are subsequently incinerated' (EC, 2000); by default, all such plants count as disposal. The criteria exempt thermal treatments which provide true recovery, where the gases 'are purified to such an extent that they are no longer a waste prior to their incineration and they can cause emissions no higher than those resulting from the burning of natural gas' (EC, 2010: 39). Although gasification had this aim, it was rarely fulfilled in practice.

For a thermal plant to be a recovery operation, it also must generate sufficient energy to fulfil the 0.65 efficiency threshold. This is calculated with the R1 formula, which relates the feedstock's calorific value to the net energy produced as electricity and/or heat, though it is not an index of energy efficiency. Below the threshold, a plant is classified as disposal (EC, 2008).

R1 classification has been mandatory for some national and local authorities to support a new EfW facility, for example, with planning permission or finance. To be eligible for Wales' subsidy of gate fees, a plant must have R1 status and be compatible with Combined Heat and Power (Welsh Government, 2012: 228). At least one proposed gasification plant was refused permission, for example, in St Helens, partly on grounds that it did not fulfil the R1 criteria (Planning Inspectorate, 2015; UKWIN, 2015). Despite sub-optimal resource recovery, many electricity-only EfW plants have gained R1 classification (Kaminski, 2015; Goulding, 2016).

Although ATTs have carried the promise of better energy recovery, such fulfilment has been rare for several reasons. Tars make the syngas unsuitable for any external use, so it is generally combusted on-site, thus counting as disposal. Reliable operation depends on extra energy input to pre-treat the feedstock, thus imposing a 'parasitic load' and reducing the net energy output relative to incinerators. Given this energy loss, better recovery depends on significant heat use, for example, via a district heating system, which cannot be easily retrofitted.

Such deficiencies in plant performance and regulatory scrutiny gave anti-incineration campaigners extra arguments to oppose new plants for merely disposing waste:

> Chapter 5 also makes clear the general unacceptability of incinerators that fail to meet the Waste Framework Directive definition of Recovery. This is bad news for the large number of Gasification and Pyrolysis plants currently proposed because they would be so inefficient that they would not meet the unambitious R1 Formula Threshold. The EfW Guide makes clear that incinerators are Disposal unless demonstrated otherwise, placing them at the bottom tier of the Waste Management Hierarchy. (UKWIN, 2014)

As such critics have argued, optimistic expectations were repeatedly shifted to hypothetical technoscientific improvements:

> Where gasification or pyrolysis facilities have been attempted, they have either failed to live up to these promises or have been suspiciously quiet about reporting their actual performance. Proponents of such technologies appear to lurch from one supposedly 'sure bet' technology to another, leaving a trail of failures and bankruptcies in their wake. (Personal communication, UKWIN, 20 March 2016)

After many years of criticism of DEFRA's policy framework, it eventually undertook to 'ensure that all future EfW plants achieve recovery status' (DEFRA, 2018: 37; see Figure 5.4). This implied assistance but also a requirement, which ATTs had rarely fulfilled.

In such ways, the opposition campaign extended the incineration controversy to 'advanced' treatments. This in turn provided a greater opportunity to counterpose a circular economy including alternative waste treatments, as explained next.

Disputing incineration: decarbonization and resource objectives

EfW policy had a dual commitment to decarbonization and resource recovery. An expert controversy arose over whether incineration facilitates those aims or rather impedes them. Building on local protests, the anti-incineration campaign intensified controversy through climate activists, local authorities and Parliament. Mainstream views eventually shifted against new incineration capacity, while counterposing alternatives which would truly

Figure 5.4: Waste hierarchy diagram clarifying that 'resource recovery' must fulfil the R1 criteria

Evolution of Waste Management Practices: In the past, most waste was dealt with by disposal, but over time that will shift incresingly to recycling, reuse and ultimately prevention.

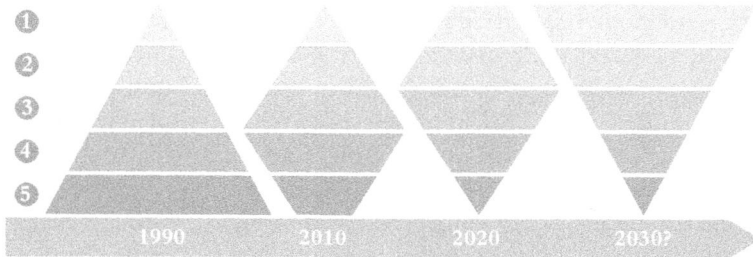

① Prevention

Using less material in design and manufacture. Keeping products for longer; reuse. Using less hazardous materials.

② Preparing for reuse

Checking, cleaning, repairing, refurbishing, whole items or spare parts.

③ Recycling

Turning waste into a new substance or product. Includes composting if it meets quality protocols.

④ Other recovery

Includes anaerobic digestion, incineration with energy recovery, gasification and pyrolysis which produce energy (fuels, heat and power) and materials from waste; some backfilling.

⑤ Disposal

Landfill and incineration without energy recovery.

Source: DEFRA (2018, revised from 2011a)

minimize or recover waste. These issues were given rival framings, each promoting a different societal future, as shown here.

More incineration plants – or a circular economy?

A decade after the PFI incineration programme began in 2006, the UK 'waste crisis' was becoming a more contested, prominent issue. Each side made contrary assumptions about the future need versus capacity for incinerating MSW into energy, while citing contrary evidence, which itself became contentious. Such disputes arose around national policy as well as in local authority decisions over new plants, which might create over-capacity for the available waste in the next decades (Reece, 2013).

As the dominant agenda, the waste-management industry repeatedly warned that the UK faced a shortage of incineration capacity to deal with rising waste. It would otherwise be sent to landfill and increase GHG emissions. 'The consensus position on waste treatment is that we will end up over five million tonnes short of energy from waste capacity by 2030. This is what the government needs to understand if it is not to sleepwalk into a capacity crisis', stated the Director of the Environmental Services Association (Creech, 2017). This claim cited evidence in an expert report (Tolvik, 2017).

Moreover, the industry argued, greater incineration capacity would be necessary for moving towards a circular economy: 'Where high quality recycling is not possible, we need options for treating residual waste according to the principles of the Circular Economy and waste hierarchy, i.e. putting waste to best use by improving energy security and contributing to economic growth' (ESA, 2018: 17). This rationale evaded the question of why so much waste was not recycled or easily recyclable, thus narrowing industry's task to energy recovery from waste.

By contrast, civil society groups criticized thermal-treatment plants for perpetuating wider systems that generate waste problems, allowing resources to be wasted and undermining recycling. For its critical interventions, UKWIN brought together various expertise – on waste, environmental economics and technological design; it cited other expert reports questioning the need for more incinerators. Its alternative agenda advocated a shift to a circular economy, which would redesign production systems in order to reduce GHG emissions and avoid waste outputs, rather than rely on waste-conversion technologies.

A similar argument was substantiated by expert reports. According to the Eunomia consultancy, EfW plants were expanding inexorably 'towards the point where we have more residual waste treatment capacity than we need' (cited in Creech, 2017). Although significant feedstock was being exported to continental incinerators, the surplus would turn into a shortage if all planned EfW facilities were built and operated at full capacity: 'together they would limit the UK's recycling rate to no more than 63%' (Eunomia, 2017). These arguments cited the European Commission's warnings that extra EfW capacity could undermine a circular economy. Public policies should avoid 'the creation of infrastructural barriers to the achievement of higher recycling rates', likewise avoid 'the risk of stranded assets' (EC, 2017: 6).

Responding to the controversy, the UK government's expert advisors warned that current financial incentives and available infrastructure may perpetuate waste problems rather than help to achieve resource productivity. Behaviour may not improve 'if the infrastructure to achieve this is unavailable', argued the UK Government Chief Scientific Adviser (2016: 25, 31). The warning related to the entire economy, though the specific relevance for incineration soon became explicit.

At a parliamentary hearing, DEFRA's scientific advisor criticized the PFI incineration programme for locking in a wasteful system through 30-year contracts and likewise criticized the government's financial incentives for aggravating waste problems: '[T]he market pull on waste encourages the production of residual waste. It encourages people to think that we can throw what could be potentially valuable materials – if we were to think about them innovatively – into a furnace and burn them' (Professor Ian Boyd, cited in Hayns-W, 2018). By investing heavily in waste incineration, Sweden had encouraged the production of residual waste, he added. In effect he was criticizing the techno-market fix of the UK policy framework, whose financial incentives undermined their official objective.

Through controversy over techno-market fixes, anti-incineration activists found extra opportunities to counterpose a circular economy. In many policy arenas, they made interventions to shift the mainstream agenda towards 'supporting sustainable practices built not on waste management but on resource management principles' (UKWIN, 2018a: 1–2). Although similar policy documents already existed (for example, EAC, 2014), by 2018 a circular economy agenda was gaining wider support as a basis to oppose extra incineration capacity.

Meanwhile Europe-wide environmentalist networks criticized the EU's ecomodernist agenda, based on 'the gospel of eco-efficiency' (cf Martinez-Alier, 2002). Instead:

> We urgently need a globally just division of resource access within environmental limits. This requires dematerialisation (factor 10), de-fossilisation (phasing out fossil fuel use by 2030), ending landfill import, and safeguarding the commons. This is much more than a circular economy and ecological modernisation. ... It is necessary to implement, in parallel, a variety of policy changes to collectively transform consumption and production patterns, legal frameworks, financial instruments, and individual behaviour. (FoEE, 2018: 36)

Rather than the ecomodernist agenda, the report advocated 'the sufficiency principle'. This alternative became a reference point for UK anti-incineration campaigners.

Climate controversy expands the opposition network

The anti-incineration campaign gained more support by engaging with climate issues and activists. The campaign's report on climate issues argued that incineration impedes decarbonization of the electricity sector:

> Electricity generated by waste incineration has significantly higher adverse climate change impacts than electricity generated through the conventional use of fossil fuels such as gas. Over its lifetime, a typical waste incinerator built in 2020 would release the equivalent of around 1.6 million tonnes of CO_2 more than sending the same waste to landfill. Even when electricity generation is taken into account, each tonne of

plastic burned at that incinerator would result in the release of around 1.43 tonnes of fossil CO2. Due to the progressive decarbonisation of the electricity supply, incinerators built after 2020 would have a relatively greater adverse climate change impact. ... Composition analysis indicates that much of what is currently used as incinerator feedstock could be recycled or composted, and this would result in carbon savings and other environmental benefits. (UKWIN, 2018b: 1)

As the policy context, the 2008 Climate Change Act had required the UK to achieve at least a 80 per cent reduction from 1990 levels; in 2019 this target was tightened to net-zero carbon by 2050. The campaign emphasized the conflict between that target and incineration when asking supporters to contact their MPs through the iParl app. By 2021 more than 5,000 messages had been sent (see the following section on the Westminster debate).

The incineration lobby was now losing the argument about incinerators saving GHG emissions relative to landfill. It justified new plants as a 'transitional technology' towards reducing residual MSW and eventually reaching net-zero carbon. A pro-incineration report blurred the distinction between various forms of EfW and incineration: 'the UK can do more to further decarbonise EfW, by getting fossil-based plastics out of the residual waste stream, and with government support to explore new carbon capture and storage technology' (Policy Connect, 2020: 7).

CCS had been anticipated as a means to reduce emissions from incinerators, though this prospect was losing credibility. The report also proposed 'that the UK should move towards a Scandinavian style approach to residual waste', that is, by increasing incineration capacity (Policy Connect, 2020: 7). According to a Green Alliance expert, however, Scandinavians now call their waste incineration programme 'crazy'; they have sought instead to reduce and recycle waste (Peake, 2020).

By 2019 many XR activists were joining local anti-incineration protests (for example, see the Gloucestershire and Camden case studies in this chapter). Several local groups jointly proposed alternative action plans for waste authorities (XR, 2020a, 2020b). They circulated their intervention strategies via WhatsApp groups.

Soon naming themselves XR Zero Waste (Figure 5.5), the network gained broad support for an Open Letter. This warned that the great expansion

Figure 5.5: Extinction Rebellion Zero Waste (XRZW) logo

of carbon-intensive EfW incineration plants was rapidly increasing carbon emissions. The UK government should make 'a world-leading transformation of its waste and resource sector', especially 'to accelerate the transition towards a genuine, zero-waste circular economy'. Such action would be necessary 'to meet our legally binding carbon budgets and to facilitate a green recovery' (XRZW, 2020). This agenda gained broad support from diverse civil society groups, prominent politicians and numerous experts.

The joint Open Letter appended an ambitious action platform, 'Transforming the UK's waste and resource sector: a blueprint for regulatory reform and structural investment'. As a major obstacle, the EfW programme 'inhibits the full decarbonisation of the power sector', especially now that greater renewable sources had helped the National Grid to decarbonize its operations. The platform called for mobilizing investment in expert capacities and resource-reuse technologies already available in the UK (XRZW, 2020).

The Letter's demands became the focus for national Days of Action opposing all new incinerators. The base was broadened to more environmental and anti-racist groups. Black Lives Matter added their voices against the air pollution from incinerators, which have been disproportionately sited in ethnic-minority areas (Roy, 2020; Eminton, 2021). According to the local spokesperson for Black Lives Matter, 'We need to be calling this what it is: racism. These industries know that when they place an incinerator in an area like Edmonton' (cited in Greenpeace and Runnymede Trust, 2022: 57).

This comment denounced a proposal by the North London Waste Authority (NLWA) to replace the Edmonton incinerator with yet another one, euphemistically called a 'clean, green' Energy Recovery Facility. There ensued disputes over resource recovery and GHG savings. Proponents claimed that 'low-carbon heat' would come from the new incinerator's North London Heat and Power Project. This claim was simply repeated by politicians who had responsibility for the NLWA.

The 'Stop the Edmonton Incinerator' campaign sought to discredit those claims about resource recovery. According to such critics, the incinerator would create excess capacity, would undermine the government's 65 per cent recycling target, would commit the NLWA to a long-term waste supply, would burn 150,000 tonnes of plastic per year would generate 700,000 tonnes of GHG emissions per year for decades; hence it would undermine efforts towards a circular economy. The campaign was joined by a coalition of seven local XR groups. Together they intervened in the local authorities which comprise the NLWA.

They advocated practical alternatives for greater resource recovery, including more recycling at several stages. In particular, a state-of-the-art materials recovery facility (MRF) could extract substantial recyclables from MSW; this would reduce residual waste, the need for incineration and thus GHG emissions. Promoting those alternatives, their joint statement declared: 'If all seven north London councils were to implement similar action plans, the combined tonnage of residual waste they send to incineration would drop by more than 50%' (XRZW, 2021). Several waste management action points would 'help Camden Council cut its non-recyclable waste by 65% and reach at least 70% recycling by 2030'. This alternative agenda strengthened the campaign against a new incinerator and against local authorities' support for it.

For several years UKWIN had been contributing to the international environmental alliance GAIA. It published a report on how zero waste systems create more jobs, create better jobs, higher wages, more permanent positions and improve quality of life – far more than incinerators. Such systems are also necessary to reduce GHG emissions. If London were to recycle or compost 80 per cent of the recyclable and organic material in its waste stream, then the city could create around 5,000 new jobs (Ribeiro-Broomhead and Tangri, 2021: 4, 14).

MPs speak against incineration, for a circular economy

For several years, local anti-incineration campaigns had been asking their MPs to convey their protest and to support alternatives. MPs increasingly reiterated such arguments in parliament. Moreover, they held high-profile debates in Westminster Central Hall in 2020 and again in 2021. Using the iParl app, more than a thousand people sent a message asking their MP to attend.

At the 2020 parliamentary event, 14 MPs gave anti-incineration speeches. They called for an incineration tax, a moratorium on new incinerators, more powers for the Environment Agency to better regulate them, tighter emissions standards and a greater push for recycling. They criticized pro-incineration arguments as 'industry greenwash' contradicting the waste hierarchy, net-zero carbon and a circular economy. MPs praised anti-incineration campaigners for dealing well with such technically complex issues (Hansard, 2020; UKWIN, 2020).

Numerous MPs opposed gasification plants as incineration, regardless of claims for 'advanced technology'. One testified:

> In my 15 years of being an MP, no other issue has galvanised so many people and brought them together against something in the way this issue has. It really is a community movement, with campaign groups such as No Monster Incinerator in Washington or Washington and Wearside Against Gasification leading the way to oppose the application by informing local residents and getting signatures on petitions. As I mentioned, 10,800 people have so far signed a petition in opposition, which I presented to Parliament last month. (Hansard, 2020; quoted in UKWIN, 2020)

At the 2021 follow-up debate, MPs spoke even more strongly against incineration, linking social justice and environmental issues. Some cited a report that waste incinerators are three times more likely to be built in the UK's most deprived neighbourhoods than in the least, bringing multiple environmental nuisances (Unearthed, 2020). One MP represented the Labour Party stance as follows:

> We should now acknowledge that the time for incineration is over and that the age of incinerators should come to an end. ... That over-reliance will prevent us from moving up the waste hierarchy in dealing with waste generally and will stop us looking at waste as a resource that can be recycled and reused, its value unlocked rather than buried or contributing to toxic air. (Daniel Zeichner MP, quoted in UKWIN, 2021)

Those speeches dismissed technofix promises. For at least a decade, incineration promoters had anticipated that such plants would eventually include a CCS facility in order to justify them as low carbon. The statutory Committee on Climate Change had recommended such a requirement for all new plants by 2020. Yet CCS still remained a distant hypothetical prospect. Some MPs cited this gap as a reason to oppose new incineration plants. According to an MP, 'We have no time to invest in low-carbon technologies; we need to put all our efforts into net zero solutions' (UKWIN, 2021). This became an extra reason to levy an incineration tax.

Conclusions

The chapter began with the following questions:

- How were market-type incentives meant to stimulate novel technologies fulfilling environmental policy aims, especially bringing waste up the hierarchy and saving GHG emissions?
- How was techno-innovation meant to go beyond incineration and overcome its limitations?
- How did protest dispute those claims and use the opportunity for alternative agendas?
- How does this conflict relate to system change versus continuity?

As the context of the UK's waste-management policies, the EU has promoted the waste hierarchy for valuing waste as a resource, alongside requirements to reduce landfill. Local authorities were meant to outsource solutions through competitive bidding for private-sector contracts which could bring improvements. In parallel the EU's innovation framework sought to incentivize eco-efficient technoscientific solutions, expecting to reconcile economic growth with environmental objectives such as resource recovery and GHG savings.

Adapting the EU waste hierarchy, the UK government devised a techno-market framework. The UK introduced several market-type measures, for example, landfill taxes, renewable energy quotas and electricity subsidies. Together these were meant to incentivize better waste-conversion techniques at the interface of waste and energy issues. In its sociotechnical imaginary, such incentives would stimulate eco-efficient conversion processes, outputs substituting for fossil fuels and GHG savings, while bringing waste up the hierarchy. As a major investment to reduce landfill, the government financed the EfW programme, subsidizing numerous new incinerators. These provoked widespread protest and sometimes deterred more incinerators.

Meanwhile a joint state–industry agenda promoted ATTs, meant to bring several improvements. From an STS co-production perspective, this sociotechnical imaginary has co-produced complementary forms of three components: resources (raw materials mainly for energy recovery, anywhere), technoscience (ATTs smoothly accommodating heterogeneous feedstock, recovering their energy value and contributing to decarbonization), and socio-political order (market competition for contracts, driven by renewable electricity subsidy). State support measures would facilitate technoscientific advance in thermal treatments, going beyond incineration and its disadvantages (see Table 5.1).

Despite those techno-optimistic promises, many local protests opposed ATTs as well as existing plants on health and nuisance grounds, especially when new plants malfunctioned. Given their siting in low-income areas, this expansion broadened the meaning of 'frontline communities'. The wider anti-incineration campaign criticized all thermal treatments for wasting resources, criticized lax standards for resource recovery and stigmatized ATTs as 'incinerators in disguise'. More industry experts questioned industry claims that ATTs would control health hazards, bring waste up the hierarchy and take waste-conversion beyond disposal. More voices criticized the linear economy that systematically perpetuates waste, facilitated by incinerators. The demand for 'a circular versus linear economy' has served as bridging frame by aligning otherwise disparate criticisms of incinerators.

As numerous ATTs underwent technical and financial failure, moreover, the waste-management industry had sharper internal disagreements over technological promises and appropriate financial incentives. Citing these disagreements, opponents sought to discredit all thermal treatments for wasting resources. As a mobilized counter-public, the anti-incineration network linked activists with diverse experts, local authority officials and politicians. Some criticized 'industry greenwash'; any expansion would contradict the waste hierarchy, net-zero carbon and a circular economy.

Table 5.1: Rival sociotechnical imaginaries of a low-carbon waste-energy future

Rival imaginaries	Techno-market fix	Eco-localization
Supporters	Large energy and waste-management companies; technology suppliers.	*Zero Carbon Britain*, Campaign Against Climate Change, Friends of the Earth, UKWIN.
Responsibility assigned	Waste authorities outsource waste management to companies judging the market advantage of various technological options, given the financial incentives.	Waste authorities should promote source segregation, minimize waste flows, optimize output usages and thus localize responsibility for the waste hierarchy.
Public good	Low-carbon technologies will more efficiently convert feedstock as low-carbon inputs for centralized systems, thus greening them.	Low-carbon circular systems should minimize resource burdens, localize resource flows and diversify output uses.
Co-producing: Resources Technoscience Socio-political order	Resource recovery emphasizes inputs for energy production. ATTs go beyond incineration, providing high-value outputs and so bringing eco-efficiency gains. Financial incentives for market competition will drive resource-optimal thermal treatments.	Resources should be conserved for reuse, towards a world without waste burdens. Process design with MRFs can enhance resource reuse and productivity. State support measures should localize institutional responsibility and democratic accountability.
MSW thermal treatments (incinerators)	Their feedstock reduces landfill burdens. Outputs can contribute to gas or electricity grids anywhere (as a global good), substituting for fossil fuels.	Incineration perpetuates the linear economy, wasting resources. System redesign should minimize waste and make it more re-usable, thus making flows circular and raising resource productivity. (This contrast served as a bridging frame aligning diverse frames.)
Localization of waste management	Small-scale, more eco-efficient thermal treatment technologies will facilitate means to localize waste management.	Local authorities should localize waste management for a circular economy. Production systems should be redesigned to minimize waste and its transport.
Circular economy and net zero	Thermal treatments more efficiently convert waste that cannot be recycled, thus providing a transitional means towards a circular economy and net zero.	Thermal treatments lock in local authorities to long-term contracts increasing demand for waste feedstock, thus impeding the aims of a circular economy and net zero.

Some MPs echoed campaigners' arguments for alternatives valuing waste as a resource, rather than awaiting future low-carbon technologies.

Underlying the anti-incineration campaign has been an eco-localization imaginary. This has co-produced distinctive forms of three components: resources (reuse by design, waste reduction), technoscience (MRFs, greater resource productivity, circular economy) and socio-political order (local responsibility for resource usage, sufficiency principle). Their campaign reinforced widespread doubts about whether the UK policy framework could achieve its environmental aims through thermal treatments. This agenda contrasts with greenwashing incineration through MRFs, as in controversies over new facilities in Gloucestershire and Edmonton.

Together opponents put the incineration industry on the defensive for wasting resources. Public controversy was extended from specific plants to the entire incineration programme, its state incentives and its dubious promises. Oppositional arguments were taken up by campaign groups coming from diverse issues, for example, health, environmental, anti-racist, and so on. Indeed, the incinerator programme extends colonial legacies to environmental racism at home, thus worsening climate injustice and stimulating more opponents. Eventually a broad network was opposing new incineration capacity as a multiple threat, while counterposing resource-recovery facilities towards a future circular economy. By 2021 opponents had deterred local waste authorities from approving more than 70 new incinerators, though some others went ahead.

Amidst public controversy, then, rival agendas have contended for influence through divergent sociotechnical imaginaries, innovative designs and expertise for them. These rival agendas gave divergent meanings to resource recovery, GHG, savings and circular economy (see Table 5.1). Although the UK policy framework had mentioned circular economy, these agendas could remain as merely technical-administrative proposals or as better resource recovery from waste. Through the anti-incineration campaign, the concept of circular economy gained political impetus and a transformative potential, as a step towards system change.

6

Green New Deal Agendas: System Change versus Continuity

Introduction

The 'Green New Deal' concept has been widely taken up for envisaging a low-carbon, environmentally sustainable, socially just future and for building alliances which could realize it. Such agendas remain largely efforts towards a deal, that is, an institutional agreement that could be implemented with substantial resources. Hence this chapter will analyse various GND agendas rather than deals per se.

Extending the GND concept, the Climate Justice movement has generated various proposals for a Global GND. These emphasize North–South injustices, their drivers and means to overcome them. Although such proposals remain even more distant from a deal, they sharpen social justice issues for national agendas as well as global ones.

Advocating a Global GND, the UK-based campaign group War on Want has drawn on its long-time role in building solidarity with 'frontline communities'. Beyond being simply a victim, such communities often confront the forces responsible for resource plunder and degradation. As the campaign has argued, a Global GND is necessary 'to deliver climate justice, by reordering our economies to protect both people and the planet'. Both face continuous threats from neocolonial extractivism. 'We must cut emissions fairly, and move away from systems of limitless extraction and exploitation; which line the pockets of corporates and rich shareholders, whilst keeping the majority of the world's population in poverty' (WoW, 2021). Such proposals highlight North–South socio-economic inequities, which remain ambiguous or obscure in North-based national versions of a GND.

Another version of a Global GND, bringing together legislators worldwide, likewise highlights North–South inequities. In particular, '[f]ood and water security are increasingly serious issues that disproportionately impact those in the Global South, exacerbated by an industrial food system that is over reliant on fossil fuels and which contributes significantly to climate change through land clearing'. Hence a Global GND must restructure economies North and South (350.org, 2021). Likewise from a climate justice perspective, a *People's GND* proposes infrastructural-agricultural transformation in the global North, and a North–South industrial convergence towards overcoming inequities (Ajl, 2021).

Numerous North-based organizations have co-sponsored *The Green New Deal for Europe* (GNDE). This seeks 'accountability for Europe's historic role in resource extraction in the Global South', as a basis for funding frontline communities to deal with colonial legacies as well as climate change. It acknowledges European people's disparate concerns and interests that may impede an alliance, for example, coal-dependent versus low-carbon regions. By first acknowledging such differences, a transnational movement can bring together diverse groups to 'stand together behind a single policy vision' (GNDE, 2019: 11, 21).

Moreover, the GNDE agenda highlights the EU's persistent commitment to neoliberal policies, as in the European Commission's Green Deal (cited in Chapter 1). For example, the EU 'subsidises private investors, socialising the risks of the green transition, while privatising the gains'. The response 'must be a logic of confrontation, pitting Europe's communities against Europe's institutions that seem unwilling to see the climate and environmental crisis through the lens of their lived realities …' (GNDE, 2019: 20, 21). Indeed, for frontline communities in the global North as well as the South, the EU's neoliberal policies worsen harm from climate change.

Hence a confrontational strategy is necessary to oppose those policies, especially the marketization, enclosure and plunder of resources. An effective alliance would need strong support from a political party that could lead and implement a GND, overcoming contrary pressures on the state apparatus. But how? 'A properly political critique would contend that the Green New Deal sustains the fantasy that an enlightened state can save us from climate catastrophe, a fantasy that discourages us from taking the radical actions that are, in fact, a prerequisite for the state doing anything at all' (Riofrancos, 2019).

Of course, the Climate Justice movement already was building radical actions to pressurize the state and business. A socially just, low-carbon transition would depend on civil society revolt strengthening working-class demands along similar lines (Beuret, 2019: 5). Greater political strength depends on overcoming or avoiding divisions within the labour movement, while also avoiding system continuity through deceptive climate fixes.

Such an effort in the UK is the main focus of this chapter. The analysis focuses on GND agendas as means to identify potential strategies and obstacles. It will explore the following questions:

- How did GND agendas address decarbonization in diverse ways, especially for high-carbon sectors?
- How did each agenda envisage societal futures? In ways differently shaping Green and Deal? What conflicts arose between those agendas?
- How did such issues arise in GND local initiatives to decarbonize heat in buildings?
- How do these various agendas relate to system change versus continuity?

As shown here, long-time political struggles have shaped the prevalent GND agenda; likewise forms of Green and Deal, meaning forms of social order underlying a potential deal. Amidst trade-union divisions, a cross-class partnership has been promoting high-carbon continuity through climate fixes. This partnership can be undermined by counter-publics building collective self-confidence for a socially just, low-carbon future. This potential has arisen firstly in some urban contexts.

This chapter has the following structure: how Green New Deal agendas entail political divergences, potentially softened through technofixes for high-carbon systems, how such tensions have played out in US Democratic Party and UK Labour Party agendas for a GND; how political divergences have arisen in local agendas to decarbonize heat by retrofitting houses; and, finally, the concluding section returns to the questions posed above.

Green New Deal agendas: contentious transition models

The GND concept has a long history, dating from US and UK responses to the 2008 financial crisis. About a decade later, the concept was revived by Left-wing forces within the labour movement. These GND agendas inspired widespread debates over a transition to a low-carbon economy. They generally neglect North–South injustices, as highlighted in proposals for a Global GND; likewise they neglect ongoing environmental degradation to extract rare earths for storing renewable energy (Riofrancos, 2020, 2022; IEA, 2021c; OPSAL, 2021).

Even within the narrow scope of national GNDs, low-carbon transition models have become politically contentious. Political divergences have arisen implicitly in sociotechnical imaginaries underlying GND agendas. Before examining them, let us survey some generic tensions around decarbonization fixes.

Tensions over decarbonization fixes

The GND concept has resonances with the 1930s US New Deal, where a rising labour movement transformed the political-economic system. The previous laissez-faire anti-working class regime was turned into a social-democratic one, expanding infrastructural investment as the basis for a higher-wage, higher-productivity economy. Its legal guarantees for workers' rights facilitated mass membership across industry-wide trade unions, in turn benefiting workers' collective power in general.

Those resonances with workers' collective agency have helped build widespread support for a GND today. In the US and UK, GND agendas promise employment security. Early versions inspired collective imaginations about lower-energy, low-carbon systems with secure high-quality trade-union jobs. Here lies a tension: While some GND agendas propose to decarbonize material production, others emphasize the need to expand labour-intensive caring activities. These have several aims, for example, to valorize current skills, to provide socially useful employment and to minimize burdens on natural resources (Bhattacharya, 2019).

System change would depend on a political commitment for significant replacements, in particular: public goods replacing many individual consumer goods, cheaper renewable energy replacing fossil fuels, low-carbon public transport replacing some private transport, agroecological methods replacing industrial food production, a plant-based diet replacing meat, and so on. Structural changes could reduce today's energy usage by 40 per cent by the year 2050, thus facilitating decarbonization of the remainder (Grubler et al, 2018). As regards such system change, prevalent transition agendas remain ambiguous: sociotechnical imaginaries envisage decarbonization technofixes, which can facilitate high-carbon system continuity.

For several decades, elite agendas have envisaged 'clean' fossil fuels as a long-term future, offering a politically convenient evasion of responsibility, as highlighted by critics (Greenpeace USA, 2015; Smolker, 2015; Krüger, 2017). In capitalist agendas for Green Growth, technological-managerial innovations would somehow reconcile climate protection with economic growth. These agendas anticipate that technofixes will eventually decarbonize or displace carbon-intensive systems, thanks to market forces (Dale, 2016).

The foremost fix has been CCS. For the EU's decarbonization agenda, hydrogen from fossil gas with CCS was labelled as 'clean, low carbon' blue hydrogen, before any plausible evidence that it is feasible or low-carbon (CEO, 2020). Much of Europe has locked in natural gas, so the industry has sought to ensure that this dominates any future hydrogen market. This blue hydrogen 'carries the potential to perpetuate current capital accumulation practices that rely on the exploitation of our natural resources in an unsustainable manner' (Szabo, 2021: 105). As an alternative pathway, green hydrogen from electrolysing water has a low-carbon potential, whose fulfilment would depend on real-time public control over energy generation sources (TUED, 2022: 51–52).

Such future technofixes play an everyday role in system continuity, regardless of whether they ever become commercially viable or low-carbon. In 2017 the International Energy Agency regretfully classified CCS as 'not on track'. Nevertheless optimistic expectations prevailed: its net-zero carbon scenario for the year 2060 included 7 per cent energy from fossil fuels with CCS, as well as 2 per cent from biomass with Bio-Energy CCS (IEA, 2017). An update reported more progress, but with scant detail on CCS turning fossil fuels into hydrogen and no cost information. As an overview of installation plans, CCS was 'not on track to reach the Net Zero Emissions by 2050 Scenario' (IEA, 2021d).

Numerous expert reports have emphasized the limitations of CCS. It needs a great energy input, undermining the potential savings in GHGs; it captures only some of the CO_2 emissions from any installation (Jacobson, 2020); and it has been effective in the dual economic-technical sense only through Enhanced Oil Recovery, that is, pumping the carbon into an oil reservoir to increase the pressure and so facilitate extraction (GTM, 2018; also Schlissel and Wamsted, 2018). That kind of success worsens the climate problem. Apart from that exception, US investors have abandoned many CCS pilot projects. CCS remains implausible as a significant means to decarbonize fossil fuels in the foreseeable future, if ever.

Such expert criticisms have been taken up especially by public-sector unions, which are prominent in Trade Unions for Energy Democracy. Its reports have emphasized the fundamental problem, namely: capitalist economies have driven rises in resource usage including fossil fuels. And CCS has minimal prospects to mitigate their GHG emissions. In the US context of the coal industry: 'even if CCS is deployed on a mass scale, the health impacts and environmental damage associated with extracting, transporting, and burning coal will not be eliminated and may become worse due to the "energy penalty" associated with CCS' (TUED, 2015: 2). CCS likewise remains problematic for decarbonizing natural gas. CCS techno-optimism has been driven by a search for economic competitiveness and growth. Instead, trade unions must 'develop transformational strategies that are anchored in a paradigm of sharing, solidarity, and sufficiency' (TUED, 2018: 43).

Social scientists have reinforced those doubts about CCS, critically analysing the wishful basis for optimistic expectations (Lohmann, 2009; Markusson et al, 2017). Even a modest financial investment serves to reconcile 'low carbon' scenarios with fossil fuels, especially coal in the US and natural gas in the UK. Through such technofixes, dominant sociotechnical imaginaries potentially accommodate trade unions in carbon-intensive sectors, their companies and finance capital.

Such imaginaries help to greenwash mainstream capitalist agendas for Green Growth (Dale, 2016). This has been defined by advocates as follows: 'Green growth means fostering economic growth and development, while ensuring that natural assets continue to provide the resources and environmental services on which our well-being relies' (OECD, 2011: 6). Even if new energy sources are truly low carbon, they often supplement fossil fuels rather than replace them, given the trend of rising energy usage. Those contradictory agendas provide the wider context for GND agendas, especially how their low-carbon transition would decarbonize energy sectors. Let us examine how political divergences have arisen around GND agendas in the US and then the UK.

US Green New Deal agendas

In the US a GND agenda originated in the Sunrise Movement. In 2018 it declared: 'The Green New Deal is a 10-year plan to mobilize every aspect of American society to 100% clean and renewable energy by 2030, a guaranteed living-wage job for anyone who needs one, and a just transition for both workers and frontline communities.' It sought 'to end the corrupting influence of fossil fuel executives on our politics' (Sunrise Movement, 2018). It opposed the neoliberal politics dominating the Democratic National Committee and its preferred candidates in primary elections.

The groups' joint demands included: 'Halt all fossil fuel leasing, phase out all fossil fuel extraction, and end fossil fuel and other dirty energy subsidies.' Sunrise's 2018 electoral successes included the new US Representative Alexandria Ocasio-Cortez, popularly known as AOC. She was already a prominent figure in the Justice Democrats, seeking to elect politicians free of corporate funding. In February 2019 she and fellow US Representative Ed Markey co-sponsored a GND House Resolution (HR), which gained numerous Congressional sponsors.

The Resolution advocated an economic transformation that would shift socio-political power through mass participation: 'A Green New Deal must be developed through transparent and inclusive consultation, collaboration, and partnership with frontline and vulnerable communities, labor unions, worker cooperatives, civil society groups, academia, and businesses.' Likewise this agenda has sought to include dispossessed groups by 'stopping current, preventing future, and repairing historic oppression of indigenous peoples, communities of color, migrant communities, deindustrialized communities, depopulated rural communities, the poor, low-income workers, women, the elderly, the unhoused, people with disabilities, and youth' (US HR GND, 2019).

Such a popular mobilization would shape public investment in 'clean green' technology, thus contributing to the following aims:

[T]o achieve net-zero greenhouse gas emissions through a fair and just transition for all communities and workers; to create millions of

good, high-wage jobs and ensure prosperity and economic security
for all people of the United States; to invest in the infrastructure and
industry of the United States to sustainably meet the challenges of the
21st century. (US HR GND, 2019)

The GND concept had enthusiastic support from many US trade unions,
especially those in the public sector, even before the February 2019 HR.
Although it promised employment security, including workers in fossil-
fuel sectors, trade unions there raised objections. The AFL-CIO Energy
Committee strongly criticized the GND HR on several grounds, for
example, that it focuses overly on decarbonizing energy production, lacks a
credible plan to decarbonize transport, omits engineering solutions such as
CCS, and so on. Thus it 'makes promises that are not achievable or realistic'
(AFL-CIO Energy Committee, 2019). Among other reasons, this divergence
within the labour movement limited prospects for a transformative GND
in the Biden Administration that began in 2021.

During the November 2020 election campaign, the Democratic Party
leadership cited the GND concept and adapted its language. 'The Biden
Plan for a Clean Energy Revolution and Environmental Justice' offered some
genuine social and environmental benefits. Yet the plan remained ambiguous
about carbon-intensive industries, whose decarbonization was left dependent
on technofixes: 'We are on the cusp of breakthroughs in technologies',
especially carbon capture, use and storage and advanced biofuels; these were
celebrated as if imminently viable. It mentioned transport solutions through
community empowerment, but promised only 'input from community
stakeholders' (Biden Climate Plan, 2020).

After the 2020 election, the new Biden Administration proposed large
expenditure incorporating some elements of a GND. According to its
original proponents, some new investments went beyond the previous
decarbonization framework assuming that 'the market is going to fix a
problem that is created by the market' (quoted in Kurtzleben, 2021). For
decarbonization, the American Jobs Plan offered financial assistance mainly
as investment and production tax credits for renewable energy developers, as
well as vouchers for consumers who trade in their internal combustion engine
cars to purchase electric vehicles (White House, 2021). The Administration
promised community and worker empowerment yet directed the industry-
relevant funds mainly to entrepreneurs, thus leaving political-economic
power in their hands.

Technoscientific investment invoked the US's global competitive
advantage: 'We can export our clean-energy technology across the globe and
create high-quality, middle-class jobs here at home' (White House, 2021).
This priority reinforced large-scale private-sector interests in the name of
the country's prosperity and decarbonization.

By contrast with the Biden Administration agenda, the Justice Democrats' GND advocated public investment as dual means for ensuring decarbonization and shifting political-economic power. Its agenda to repair and retrofit public housing would remove mould, as well as provide social amenities such as gardens, daycare, bookstores and grocery stores. Here is the crucial distinction: 'Are you giving *people* stuff or giving companies money to give people stuff?' Hence the inspiration from the original New Deal, whose initial advances strengthened working-class power as a basis to make further gains (Aronoff, 2021; see also Aronoff et al, 2019). Moreover, 'Decarbonizing the power sector and retrofitting homes and buildings to run on that new grid, as Green New Deal proposals outline, would not just create millions of jobs but decouple life's essentials from fossil-fuel price volatility' (Aronoff, 2022).

Such an advance would depend on a coalition taking away political-economic power from corporate interests. As a contrast with the Biden Administration agenda, this highlights divergent meanings of Green (for example, environmental protection, housing retrofits and public-good amenities versus simply low-carbon resource efficiency through technofixes) and Deal (working-class power versus corporate subsidy and control). GND agendas envisage and construct the future along those divergent lines. Next let us examine how analogous tensions arose within UK labour movement agendas for a GND.

The UK Labour Party's Green New Deal

In parallel with the US Sunrise Movement, the UK group 'Labour for a Green New Deal' arose from Left-wing forces in the Labour Party. Its agenda promoted a 'socialist zero carbon' economy by 2030, rather than the government's net-zero target for 2050 with great scope for carbon offsetting. The more ambitious target would enable the Labour Party to 'lead a radical reconstruction of our country from the ground up'. Its agenda would 'transform the economy through unprecedented investment in technology, infrastructure and people'. It declared, 'We can take the economy out of the control of the super rich, and put it in the hands of ordinary people' (LabGND, 2019a).

Contradictory decarbonization agendas

The GND initiative conflicted with some trade unions jointly calling themselves 'the energy unions', really meaning carbon-intensive sectors. At the previous year's TUC conference they had sponsored a motion to 'develop a political and lobbying strategy' for a just transition to a low-carbon economy, 'led by the voices and experiences of energy unions and their members'. In particular, the government should ensure 'a balanced energy mix', which would include 'investment in renewables, new nuclear and lower-carbon gas'. This agenda provoked dissent, especially from public-sector unions, which were marginalized (Farand, 2018).

That divergence was replicated within the Labour Party through rival motions. Internal negotiations resulted in the 2019 GND conference motion (LabGND, 2019b). Weaker than the original version, it was silent on several issues, for example, North Sea oil extraction and nuclear power, which the party had already supported. It undertook to 'take transport into public ownership and invest in expanded, integrated, free or affordable green public transport that connects Britain'. It promised 'a radical car scrappage scheme to increase electric vehicles', but not an effort to reduce private car transport or its energy demand.

As Labour's shadow minister for Business, Energy and Industrial Strategy, Rebecca Long-Bailey MP announced an agenda for an 'electric car revolution'. The latter emphasis came from the Labour Party's agenda for a Green Industrial Revolution (GIR), which had been developed separately from the Green New Deal agenda. Through numerous 'Transform Your Town' events, the GIR campaign built a network of activists and proposals for local economies (Labour Party, 2019a).

Those events were stimulated and coordinated by the Labour Party's Community Organizing Unit, meanwhile training hundreds of local activists. They made special efforts to engage people in deindustrialized towns, coastal areas and ex-mining areas. Each event gave prominence to climate campaigners, youth climate strikers, women and ethnic minorities. The organizers opened up the questions: What would a just transition mean? What would it mean in this community? In many places, action proposals focused on a specific local need such as a recycling centre, bus provision or care work (Josette, 2022).

By contrast to that community-building focus, the Labour Party's GIR manifesto was driven by some trade unions. In particular, Unite the Union emphasized state investment in capital-intensive projects, especially 'finance to re-shore supply chains and reduce the global carbon footprint of manufactured goods' (Unite, 2019: 11). This assumed that UK technoscientific innovation would decarbonize such production more effectively than other countries. The GIR's agenda to re-industrialize Britain shaped the 2019 election manifesto along different lines than the original GND proposal.

The election manifesto undertook 'to achieve the substantial majority of our emissions reductions by 2030 in a way that is evidence-based, just and that delivers an economy that serves the interests of the many, not the few' (Labour Party, 2019b: 13). The latter phrase, drawn from Shelley's famous poem, was the main campaign slogan. In the spirit of that slogan, Labour's GND promised transformation – but also entrenched much continuity.

As the wider political context, previous governments had privatized the energy sector and promoted dependence on natural gas. In response, the Corbyn leadership had proposed to nationalize the energy industry as necessary means to implement strong policies on climate change and socially just electricity provision. Yet its election manifesto promised to nationalize only the distribution grid, not the Big Six energy generation companies (Labour Party, 2019b: 16).

In parallel the manifesto promoted public–interest alternatives: 'We will expand distributed and community energy. ... We will support energy workers through transition and guarantee them retraining and a new, unionised job on equivalent terms and conditions' (Labour Party, 2019b: 16). A Labour government would also tax the tech giants to 'pay for the operating costs of the public full-fibre network' and would break their hold on advertising revenues (Labour Party, 2019b: 53).

A broader agenda would re-industrialize the UK economy:

> We can invest in our Green Industrial Revolution, deploying our collective resources to rapidly green our economy, creating one million skilled jobs and laying the foundations of a society that thrives and endures. ... And by taking radical action at home, UK companies will be in a prime position to make up a large share of the global green economy which by the late 2020s is expected to grow to $9 trillion. (Labour Party, 2019c: 3)

These aims identified with UK-based capital gaining a global competitive advantage through economic growth.

Likewise presuming system continuity, this agenda emphasized capitalist enterprises as the main means for greening production methods. With its GIR agenda, the Labour Party evoked features of the original Industrial Revolution – productivity, innovation, economic growth – as if these were readily compatible with social justice and environmental sustainability. As noted by a founder of Labour for a Green New Deal, 'the GIR proposes treatment for the symptoms of our ailing economic model – investment in green industry, decarbonisation of vital services like energy and transport – without proposing a cure for the disease itself' (Buller, 2020).

In the Labour Party's agenda, the necessary finance would come from a Green Transformation Fund. Its narrative contrasted past industrial failures with a globally more competitive economy, especially by anticipating hypothetical technofixes:

> Britain was left behind in the race to develop wind and solar technology: Labour won't let that happen again. Our plans will put UK companies and workers in pole position to lead the world in

designing and manufacturing the next generation of green technology, including marine power, hydrogen and carbon capture and storage (CCS). (Labour Party, 2019c: 13)

As an engineers' group warned, CCS could not make a significant contribution to decarbonization until mid-century, contrary to techno-optimistic approaches. 'The hope of an invisible, technology-led, solution to climate change is obviously attractive to politicians and incumbent businesses.' However, this deterred examining UK patterns of energy demand, which had risen in all non-industrial sectors (UK Fires, 2019: 7). To ensure that this increase comes from low-carbon sources, plans should rely on already proven technologies.

Despite those warnings, and despite the Left-wing shift towards the Corbyn leadership, the Labour Party's nationalist ecomodernist framework accepted the capitalist CCS agenda. It had continuities with New Labour's sociotechnical imaginary from a decade earlier. Both have imagined the nation as a competitive economic space to generate eco-efficient capital-intensive innovation. Hence the Labour Party's GND encompassed contradictory agendas.

System continuity through cross-class partnership

According to the Green Transformation Fund agenda, CCS would provide 'zero-emissions hydrogen' to decarbonize energy-intensive sectors and provide thousands of secure jobs, as an imperative for large state investment (Labour Party, 2019c: 13). This agenda evaded crucial questions, namely: how much energy would be necessary to extract, transport and reliably store the carbon from natural gas; what renewable energy source would power the process, and thus on what basis it would be truly low-carbon. Green technofixes were meant to reconcile contrary objectives.

The CCS techno-optimistic promises for 'zero-emissions hydrogen' came from the cross-class partnership in carbon-intensive sectors, especially trade unions in the natural gas industry. In 2018 this agenda had been launched as the *H21 North of England* project, a partnership between Northern Gas Networks (NGN) and Equinor, Norway's state oil company. The project would greatly expand Equinor's CCS process, as well as natural gas imports, to heat houses throughout northern England. The gas would be converted to hydrogen, often called 'blue hydrogen', which has 'close to zero emissions at the point of use' (NGN, 2018: 57). This agenda made several implicit assumptions, for example: that the CCS process would obtain its enormous energy input from offshore wind power, that this energy source would have no rival uses, and that gas extraction per se has no GHG emissions. All those techno-optimistic assumptions have been dubious; natural gas extraction has persistent methane emissions.

Despite those dubious assumptions, the NGN–Equinor 'decarbonization' agenda gained widespread support with promises of decarbonization and long-term jobs. By 2019 it was endorsed by trade unions in the fossil fuel sector, national infrastructure agencies and many Labour MPs in northern England. Through this cross-class partnership, Green and Deal were mutually shaped along energy-intensive capitalist lines, as ultimately embraced by the Labour Party leadership. This industry agenda would relegate 'the many' to the role of consumers and employees. By contrast with this system continuity, next let us examine some GND local agendas, especially to retrofit houses for heat decarbonization.

Green New Deal local agendas for retrofitting houses

For heating houses, the UK has depended mainly on natural gas, so its replacement has become a major focus of government policy and hence rival agendas. The policy has sought 'to encourage the market to deliver' appropriate means of decarbonizing and conserving heat. For the latter aim, it aimed to upgrade all homes 'to an Energy Performance Certificate (EPC) band C by 2035, where practical, cost effective, and affordable' (BEIS, 2018: 30). Its ambitious aims would be fulfilled through 'market-based mechanisms' for heat pumps, heat networks and supply chains (BEIS, 2021).

Based on such means, the 2020 Green Homes Grants scheme was a failure. It upgraded fewer than 50,000 homes out of the 600,000 originally envisaged and delivered a small fraction of the expected jobs. As some reasons why, it 'was overly complex and did not sufficiently address the needs of consumers and installers', according to an audit body (PAC, 2021). As more fundamental reasons, its householder-based voucher system fragmented the task; this basis precluded the collective knowledge necessary for monitoring and improving quality standards. This scheme exemplifies the market-based instruments of the UK neoliberal policy framework over several decades. Recognizing the scheme's failure, a Parliamentary Committee urged the government to 'scale up' the various markets (BEIS Committee, 2022).

By contrast with the market-based framework, since 2019 the natural gas industry had been asking the UK government for enormous investment to implement its CCS fix for decarbonizing natural gas into blue hydrogen. This was meant eventually to provide low-carbon heat using gas pipes throughout northern England. Like the 'energy unions', the UK's Trades Union Congress (TUC) had already promoted CCS as a crucial innovation for decarbonizing high-carbon sectors more generally and thus for maintaining members' employment there (TUC, 2014).

This techno-optimistic perspective complemented a political fatalism about any possible alternatives, as grounds for system continuity: 'We have the biggest number of dirty industries that produce more carbon than any other region. ... If we didn't have CCS, we are on target for losing 100,000 jobs' (TUC Yorkshire & Humber, quoted in TUC, 2020: 20). This perspective underlay the regional TUC's efforts to promote CCS for carbon-intensive industry in the Humber Estuary, as well as for biofuels at the Drax power station (TUC, 2020: 4, 19).

As a well-known alternative for decarbonizing heat, some local authorities have promoted and installed heat pumps to replace natural gas. Good-quality installation requires special skills, though the difficulties have been exaggerated by the gas industry partnership. According to its report, UK housing stock would face great disruption, costs and inadequate space, so that heat pumps would be a viable solution for only a minority of homes (EUA, 2021). This negative portrayal justified CCS-hydrogen development as the only large-scale option.

Despite the cross-class partnership for the CCS-hydrogen agenda, it became contentious within the labour movement. Some local Trades Councils formed political alliances for alternatives, often under the heading of GND or Just Transition. Such efforts across several cities were shared and highlighted at People's Summit events during Glasgow's COP26 in November 2021 (see also CACCTU, 2021: 37–48). The following section analyses such efforts in two cities, Leeds and Glasgow. Their initiatives can be understood as counter-publics contesting elite agendas and counterposing socially just, low-carbon alternatives.

Leeds Trades Council versus CCS-hydrogen agenda

Based in Leeds, the H21 North project for decarbonizing natural gas provoked some activists there to oppose the gas industry agenda. The issue was taken up by the Environment Committee of the Leeds Trades Union Council (henceforth Trades Council, to avoid confusion with the TUC and its regional affiliates). In mid-2020 several webinars explored alternative options for decarbonizing heat in houses. These events featured talks from expert and stakeholder perspectives on various aspects, for example, the CCS-hydrogen technofix, building standards for heat insulation, training workers, retrofitting techniques, government finance, future societal visions, and so on. One participant had trained some neighbours in the necessary skills to help retrofit each other's houses with high-quality insulation; this grassroots innovation was developing a knowledge commons. In the webinars, most participants opposed the CCS technofix, with some citing critical reports (for example, TUED, 2015; CEO, 2020).

Subsequently the Leeds Trades Council (2020) issued a call for action for a retrofit programme. It was meant to stimulate local government responsibility for a multi-stakeholder coalition: 'The scheme should be coordinated by Leeds City Council, in partnership with unions, practitioners, community groups and local training providers such as Leeds College of Building', which by coincidence had expertise for retrofitting houses with heat pumps.

The call emphasized the need to reduce energy demand as a crucial means to reach net-zero carbon, especially in houses:

> Converting all domestic heating and hot water to electricity from renewable sources is viable only if demand is significantly reduced. ... Insulating a home to a very high standard can reduce the need for energy input by around 80%, meaning that heat pumps using renewably produced electricity will provide adequate heat with a minimal carbon cost. (Leeds Trades Council, 2020: 4)

As advocated in various expert reports, high-standard insulation could save much money and so also address fuel poverty. Heat pumps are expected to have lower running costs, compared with a gas or hydrogen boiler with the same package of retrofits. Like a boiler, they provide hot water as well as heating the living space; in a heat wave they could be used to remove excess heat too (GBC, 2020; IPPR, 2020; PCAN, 2020).

The call rejected the gas industry's decarbonization agenda on several grounds:

> CCS is currently unproven at the necessary scale, involves a significant additional energy demand, and is unlikely in practice to achieve the necessary CO_2 capture rates. ... The hydrogen option means a massive diversion of resources away from genuine decarbonisation and into infrastructure that locks in fossil-fuel dependency for years to come whilst simultaneously reinforcing corporate control of energy infrastructure – at the expense of democratically controlled programmes that place workers and households at the centre. ... The hydrogen option cannot produce the well-paid secure jobs that we are crying out for in this double crisis of climate emergency and soaring unemployment due to Covid. (Leeds Trades Council, 2020: 6)

Its plan emphasized new arrangements for democratic accountability:

> To ensure democratic ownership and control of a mass programme of retrofit, with proper representation of both workers and householders, the programme needs to be coordinated by the Council and developed in collaboration with the unions, as well as with residents and community organisations and retrofit practitioners. ... A successful programme depends crucially on 'buy-in' from tenants and residents in all housing tenures, with retrofit ideally carried out on a whole-street or

neighbourhood group basis, potentially across a mix of tenures within that group; or on behalf of a cooperative or a self-organised group of home-owners. (Leeds Trades Council, 2020: 11)

Despite great enthusiasm for this initiative, the Leeds Trades Council had difficulty stimulating a multi-stakeholder coalition; it had little experience in doing so.

Long before those efforts, the gas industry's CCS-hydrogen fix was being promoted by the Leeds Climate Commission (2019). This fix strangely coexisted with the Commission's broader proposals, which included retrofitting houses with heat pumps (Gouldson et al, 2020). Given these contradictory agendas, the Leeds Climate Commission remained elusive as an ally for the Trades Council.

In 2021 Leeds City Council used funds from its Housing Revenue Account and a government grant to retrofit many high rise flats, especially with insulation and air-source heat pumps (Leeds City Council, 2021). The work was arranged through external private-sector contracts, and the heat pumps were imported. So the initial scheme lost the opportunity to build local capacity.

Nevertheless the Trades Council's retrofit agenda gained support from two broader campaigns. First, several campaign groups set up the Climate Emergency Community Action Programme. This is 'a partnership of local organisations to help transform Leeds into a zero carbon, nature friendly, socially just and liveable city by 2030'. In 2021 it was renamed Climate Action Leeds, linked with a wider campaign. The latter has demanded a mass retrofit programme, while opposing the city's development model 'which serves corporations at the expense of the people' (Our Future Leeds, 2021). By encompassing all climate issues, this alliance could better build support for a retrofit agenda.

Second, when the local authorities proposed to expand the Leeds-Bradford Airport, promising thousands of long-term jobs, this provoked an opposition campaign. Its own GND agenda identified a fundamental problem in the linear economy and counterposed a circular economy in a dual sense: redesigning production systems, and investing in the local supply chains (GALBA, 2021). Stable, skilled jobs could be created by various means such as retrofitting houses:

These jobs would facilitate a transition to a more circular and resilient Leeds City Region economy. They would bring about a transformation of energy generation; large scale retrofitting of the region's housing to make our homes highly energy-efficient; re-using and recycling of most of what is currently thrown away; and developing zero carbon, sustainable transport systems. (GALBA, 2021: 4)

Figure 6.1: Our Climate: Our Homes, Scottish TUC

Source: www.stuc.org.uk/

In those ways, a false solution for local employment provoked protest, public controversy and an alternative agenda along eco-localization lines.

Glasgow's house retrofit programme: rival agendas

With the slogan, 'Our Climate: Our Homes', the Scottish Trades Union Congress (STUC) led a multi-stakeholder proposal for a 'whole house retrofit' approach for decarbonizing heat in homes (see Figure 6.1). It included civil society groups such as the Poverty Alliance, Living Rent tenants' organizations and Friends of the Earth. They jointly demanded substantial funds and new state structures, especially a National Infrastructure Company and municipal energy companies. This plan would provide numerous unionized, green jobs for high-quality retrofits. These measures would be necessary to avoid the limitations and failures of the UK government's retrofit initiative (STUC, 2021).

This extended a previous proposal, coordinated by Common Weal et al (2019). It contributed to a broader plan: a 'Green New Deal for Scotland'. As regards the institutional means, 'genuine public-good private-public partnerships should be developed, but government should also intervene directly where it needs to' (Common Weal, 2019: 104). This section draws on an interview with two authors of those reports (Stuart Graham of the Glasgow TUC and Craig Dalzell of Common Weal, both on 10 March 2022).

For an adequate retrofit programme, a major obstacle has been Glasgow's neoliberal policy framework, dating from at least the 1980s. It has structured public expenditure as new markets aiming to incentivize entrepreneurialism and to attract business investment. This framework had generally diminished the decision-making capacity of the public sector (Boyle et al, 2008). Given that neoliberal framework of the local authority, its retrofitting plan soon conflicted with the labour movement agenda prioritizing the public good.

In 2021 the Scottish government funded the Glasgow City Region to retrofit homes and substitute renewable energy systems for natural gas. Sufficient for half a million houses, the funds were meant for jointly addressing fuel poverty, heat efficiency and decarbonization (Sandlands, 2021). Glasgow City Region announced an ambitious plan to retrofit the housing stock by 2032 (Glasgow City Region, 2021a). This was part of the Glasgow Green Deal, which promised many benefits such as 'ensuring a fairer and more equal economy' (Glasgow City Council, 2021).

Glasgow City Council organized a three-day event raising several challenges of a retrofit programme. According to the Council, suitable technology was already available to scale up the retrofit. But the programme would need 'collaboration between government, industry and training providers to realise Glasgow's aspiration of carbon neutrality by 2030'. An initial pilot was to retrofit the city's iconic tenement blocks. For the Low Carbon Homes agenda, the Council's experts mentioned issues such as social justice and fuel poverty (LCH, 2021). The plans for Glasgow to host COP26 in November 2021 intensified debate on decarbonization, strengthening the impetus for the government's plans and promises.

Their institutional framework posed several obstacles to a worthwhile, credible retrofit programme. A full retrofit programme may need until at least 2040. Yet the Scottish government made a firm financial commitment only for the 2021–2026 parliamentary term.

This short timescale provided a weak incentive for business investment in the necessary skills and local manufacturing capacity, whose gaps were well known (CXC, 2022). According to a retrofit programme manager, 'The existing short-term funding streams do not give businesses the long-term confidence of a multi-year pipeline of work that will encourage the acceleration and expansion of business investment in the skills of their staff and manufacturing capability' (Glasgow City Region, 2021b: 2). Under those inadequate arrangements, a retrofit programme would depend on the current long supply chains, especially imported expertise, equipment and materials. Alternatively, the government could make a commitment to create local manufacturing capability, as a crucial basis to realize the local economic and environmental benefits (Common Weal, 2019).

Energy performance standards were also weak or doubtful. Initial negotiations with contractors agreed pilot projects at a high standard of energy efficiency, such as Passivhaus in some cases (Paciaroni, 2021; Wilson, 2021). By contrast, the overall programme set a minimal standard, Energy Performance Certificate (EPC) level C. This would mean upgrading approximately half a million units that were at a lower standard.

However, this basis would not guarantee any specific standard in practice. According to some experts, the EPC is simply an administrative compliance method, not an energy efficiency measure or predictor, partly because the outcome depends on each householder's behaviour (quoted in LGHPC, 2021: 6). Whatever the modest gain, it may be superseded later by a higher standard, thus requiring an extra upgrade at a greater overall cost.

A comprehensive programme would need householders' enthusiasm, based on seeing initial retrofits visibly saving heat costs in other households. This has been one reason for a 'fabric first' approach, that is, installing effective insulation before alternative heat sources in order to maximize their benefit from the start. This is necessary but insufficient because a 'fabric first' approach prioritizes technical considerations. By contrast, 'an occupant-centred "folk first" approach may justify overcoming financial and related barriers which themselves do far more to restrict the choice of options for improving energy efficiency' (CommonWeal et al, 2019: 3).

Such barriers feature the EPC level C. Why did Glasgow's programme adopt this? As one driver, a minimal standard simplifies competitive tendering, the wider neoliberal regime which has driven the overall retrofit programme. This creates competition among organizations that instead could cooperatively raise standards on a case-by-case basis. According to a housing association, 'The funding is not being allocated strategically. Funding is still allocated through a bid process, which means that we are competing against other organisations' (quoted in LGHPC, 2021: 19). According to the relevant minister, 'innovative collective models of transition may play an important role in increasing the pace of retrofit' (Harvie, 2022). Yet any such model has been constrained by the regime of competitive tendering, which can be understood as a techno-market fix.

As a related obstacle, a competitive call favours large foreign companies with administrative capacity for the necessary documentation but perhaps minimal standards for energy efficiency. Price competition can drive down quality in practice. Competitive tendering excludes and/or fragments small local suppliers which have skills for higher standards. This arrangement deters the large-scale cooperative programme that would be necessary for the greater skills investment, mutual learning and differentiated approach to the diverse building types (Glasgow TUC and Common Weal interview, 10 March 2022). In those ways the programme's public-good benefits have been limited by Glasgow's decades-old neoliberal regime (Boyle et al, 2008; Webb, 2019).

Like many cities, Glasgow has lacked the skills for normal repair and maintenance of the housing stock, much less for new skills. So a strong incentive would be necessary for workers to learn retrofitting skills. As the labour movement proposal had said, '[s]caling up delivery poses a huge skills challenge, particularly given the large number of self-employed contractors understandably reluctant to take time out of being paid to learn new skills' (STUC, 2021: 3).

In 2021 Glasgow initiated such a training programme, but it had little take-up, for several reasons. The Scottish government's low, short-term financial commitment has provided a weak incentive for building-trades workers to take up the opportunity. As a plausible disadvantageous scenario, some workers who already had building skills (such as electricians or plumbers) could obtain retrofit training, find that the programme is short-lived and then have difficulties returning to their original trade. Such doubts have been voiced as reasons for weak interest in the training opportunity (Glasgow TUC and Common Weal interview, 10 March 2022).

Public-good alternative

In 2019 Common Weal, a Scottish 'think and do tank', had anticipated such obstacles and so proposed a comprehensive decarbonization plan. It lay within a broader *Common Home Plan*, also called a Green New Deal for Scotland. This would provide a comprehensive alternative to Glasgow's techno-market fix.

A key message of the *Common Home Plan* is:

> You're not powerless. ... We're going to show how a Green New Deal for Scotland will not just save the world but will benefit our country, our communities and you individually. Better food, better homes, better jobs, cleaner air, less waste and pollution and an economy based on repairing the things that we need rather than throwing away things that we don't need. ... By leading as an example, by coming up with the solutions and then exporting our skills and our innovations to others we can bring the word with us rather than sitting back or asking the world to stop so that we can catch up. (Source News, 2019)

Rather than try to drive economic growth, this plan promotes an economy of sufficiency. It invites people to help create the necessary innovation, which prioritizes repair, refurbishment, digital services and better resource use (Common Weal, 2019).

For decarbonizing heat in houses, the plan identified many ways to replace natural gas, reduce GHG emissions and address fuel poverty. Ideally, district heating systems would distribute surplus heat at low cost. Heat pumps would have a stronger rationale in rural areas, though also a wider relevance. By contrast, the gas industry agenda for CCS-hydrogen 'poses serious risks to the decarbonisation of Scottish energy supplies' (Common Weal et al, 2019: 8).

This had significant differences from prevalent decarbonization agendas. While they depend on future techno-solutions, this one is 'based almost entirely on current or old technology'. The dominant 'market-pricing-and-subsidy regime' would increase inequality, so this plan emphasizes public-sector responsibility (Common Weal et al, 2019: 11). This change would be necessary so that state procurement becomes a true public service, beyond the transactional-based profit-oriented economy.

Its plan outlined many inherent complexities of decarbonizing heat in houses, as stronger grounds for state-led institutional change, in particular: 'Set up a National Housing Company to retrofit all existing houses to achieve 70–90 per cent thermal efficiency. Change building regulations and invest in domestic supply chains to make almost all new construction materials in Scotland either organic or recycled' (Common Weal, 2019: 39).

It also made specific proposals for low-carbon heat sources. Given several disadvantages of electricity-based heat, the report promoted district heating systems as cheap, viable means to deliver renewable heat to homes (Common Weal, 2020b). For both those aspects, public-sector responsibility would be necessary to implement effective solutions and overcome potential obstacles, as the reports emphasized.

Those proposals had anticipated limitations that later arose in the Scottish government's 2021 retrofit programme and Glasgow City Region's role. In response, the labour movement network tried to persuade the Scottish government to establish the necessary long-term commitment, higher standards and institutional framework, including a National Infrastructure body for overall decarbonization. Likewise they tried to persuade the City Council to replace the competitive tendering regime with a more flexible basis that would facilitate more diverse bids and raise quality standards.

Although technical studies per se cannot overcome the problem, they can help identify institutional weaknesses and policy obstacles. Solutions need political changes which can drive economic change towards shorter, high-quality supply chains supporting long-term skilled livelihoods. The necessary political changes would depend on a high-profile campaign that links issues such as labour standards, housing quality, environmental protection, fuel poverty and so on.

To link such groups and issues, the campaign ScotE3 (employment, energy and environment) was already advocating climate jobs within a Just Transition framework. It was initiated by trade unionists and climate activists, 'keen to find a way of taking climate action into workplaces and working class communities'. Alongside its positive proposals are attacks on false solutions, especially a CCS-hydrogen technofix for decarbonizing fossil fuels. The latter has been a basis for the North Sea Transition Deal, endorsed by the Scottish government (Scot.E3, 2021).

Scotland has a special opportunity for a socially just, environmentally sustainable decarbonization agenda. This is partly due to its devolution arrangements, with a larger budget and greater legal powers than the UK's regional authorities. Yet the Scottish government has made false promises, such as through its neoliberal retrofit framework or hypothetical technofixes, thereby avoiding responsibility for decarbonization. By default, these promises may be accepted by a passive public – unless opposed by a strong alliance for climate justice.

As the Glasgow case shows, a Green (New) Deal has become a widespread banner, attracting divergent agendas for a retrofit programme. These promote divergent sociotechnical forms, linking technical standards with different social orders, especially a rivalry between market-competition versus community-worker cooperation. To realize the potential benefits, a public-good agenda would need to undermine and displace the dominant policy framework.

Retrofit difficulties as sociotechnical issues

From difficulties such as those described here, retrofit campaigners have identified many obstacles and challenges. Some are directly or indirectly related to neoliberal policy regimes. Here is a list with possible solutions:

- Building standards have been weak for heat insulation and energy efficiency, thus accommodating industry pressures. Tighter standards could reconcile comfort with a lower energy demand, as a prerequisite for low-carbon energy sources to be a worthwhile, feasible solution. Such standards initially should be a condition for state finance to retrofit homes. These would also set a good precedent for tightening statutory standards for new-build construction.
- For building regulations, there is a tendency for only step-wise requirements by specific deadlines. This approach would necessitate step-wise retrofitting programmes, which could be disruptive and unattractive to householders. To avoid this prospect, new buildings should have a requirement for the best feasible standard from the earliest possible date. This approach would stimulate skills development also benefiting a retrofit programme.
- Private real-estate owners often ignore tighter standards. Local authorities could impose Compulsory Purchase Orders on buildings of recalcitrant owners.

- Retrofitting skills have been barely developed in the UK. Local authorities want to train more workers but have difficulty to do so. This task needs cooperation with Further Education Colleges which have the capacity to expand such training.
- Construction industry employment has been notoriously casualized and so needs trade-union organization to ensure workers' protection. Local alliances can help promote in-sourcing and unionization.
- Local authorities depend on large contractors for large-scale projects, so they have difficulty imposing stringent conditions and auditing quality standards. Relative to many other European countries, district heating systems have been relatively less developed in the UK but have considerable potential, with a broad geographical reach. Their development would require changes in planning policy and resources by local authorities.
- After the COVID-19 pandemic, local authorities had even less funds for large-scale improvements and so became more dependent on the private sector or on state funds which have inappropriate terms. If retrofit schemes assign eligibility to property owners (such as the UK's Green Homes Grants), then this basis impedes a collective effort to ensure quality and deliver the multiple benefits.
- UK national policy has been to substitute low-carbon energy sources for fossil fuels, but this aim has several structural obstacles. For example, the National Grid sometimes pays renewable energy generators to stop producing power, thus prioritizing fossil sources and perpetuating them. A comprehensive solution would be public control of the grid, necessitating a battle against the large energy companies.

All those obstacles could limit the form, scale and public benefits of a housing retrofit programme. Although some obstacles may seem technical or legal, they are sociotechnical issues, intertwining technical design with social order and political power. Some solutions lie within an eco-localization strategy with shorter, publicly accountable supply chains integrating several environmental and economic-livelihood benefits.

Conclusion

The introduction to this chapter posed some questions about GND agendas:

- How did GND agendas address decarbonization in diverse ways, especially for high-carbon sectors?
- How did each agenda envisage societal futures? In ways differently shaping Green and Deal? What conflicts arose between those agendas?

- How did such issues arise in GND local initiatives to decarbonize heat in buildings?
- How do these various agendas relate to system change versus continuity?

The GND concept has been appropriated for political agendas promoting rival futures. Each corresponds with a sociotechnical imaginary expressing a distinctive co-production of Green (natural resources) knowledge and a Deal (social order). As shown here, such divergences have arisen within major political parties in the US and UK, as well as in local agendas to decarbonize heat by retrofitting houses. Both levels entail a conflict between system continuity and system change.

As a common pattern in the US and UK, since 2019 GND agendas have advocated a decarbonization process for a zero-carbon economy by 2030. This agenda gained significant support as a means to combine greater socio-economic equity, secure good-quality employment and an environmentally sustainable economy. The transformative potential inspired proposals for socially equitable arrangements such as more public goods, a caring economy and workers' cooperatives. This GND agenda gained support, especially from public-sector trade unions and environmentalist allies. It has resonances with eco-localization imaginaries.

However, the strong decarbonization agenda has conflicted with a Green Growth agenda promoting CCS as a means to perpetuate fossil fuels. This promise has provided an investment imperative for a dubious low-carbon remedy, or an alibi to await its feasibility, or both at once. Such techno-optimistic promises lie within a broader sociotechnical imaginary whereby capital-intensive technoscientific innovation will combine several benefits such as decarbonization, high-quality employment and technology exports. This imaginary helps to soften societal conflicts or disruptions in decarbonizing the economy, thus illustrating a general popular appeal of technofixes.

This agenda has been promoted by a hegemonic cross-class partnership between trade unions and employers in fossil-fuel sectors. This partnership has reinforced workers' dependence for their livelihoods on profit-driven employers. It has internalized the dominant sociotechnical imaginary of the nation as a unitary economic space, whose competitive advantage depends on capital-intensive technoscientific advance.

This tension played out differently in the two countries. The US Justice Democrats' GND was oriented especially towards 'frontline and vulnerable communities', that is, predominantly lower-income, ethnic minorities and communities of colour; they suffer the greatest harm from the fossil-fuel industries and road traffic. This GND agenda made no commitment to CCS to continue those industries and so was criticized by trade unions in the sector. By contrast, the UK's analogous trade unions appropriated the GND for their pro-CCS stance on decarbonizing natural gas, as adopted by the Labour Party leadership; this conflicted with the original commitment to a zero-carbon economy by 2030.

Nevertheless, the GND's original agenda has stimulated efforts at local versions, some resonating with eco-localization imaginaries. These envisage systemic changes: localizing production-consumption circuits, increasing public goods, enhancing socio-economic equity, producing for sufficiency and minimizing resource burdens (Featherstone et al, 2012; North and Longhurst, 2013). These aims have inspired initiatives beyond the capitalist framework of Green Growth.

Since 2019 civil society pressures have led many local authorities to declare a climate emergency, some have elaborated local versions of a GND. Their public-interest benefits depend on local authorities facilitating and funding cooperative enterprises, as an alternative to profit-driven companies, or at least imposing standards for labour protection and decarbonization (McInroy, 2020; Ridley, 2020). Such issues became more salient during the COVID-19 crisis, when businesses sought significant public funds to survive. Pilot projects could trial GND policies and deliver multiple social benefits (Buller, 2020).

In UK local cases here, the GND concept was appropriated for divergent means to decarbonize heat in houses. Leeds Trades Council rejected the CCS-hydrogen fix of many national trade unions, instead advocating a public-interest agenda to retrofit houses with insulation and heat pumps. For such decarbonization, the Scottish labour movement elaborated a GND public-good framework. This would stimulate cooperative knowledge-exchange for flexible, better-quality designs increasing energy performance. The process would involve grassroots innovation and a knowledge commons in skills training. Societal benefits would depend on greater state responsibility through new institutions such as an infrastructure agency. Such municipal campaigns combine several issues (for example, decarbonization, building-quality standards, fuel poverty, labour in-sourcing, and so on), thus broadening the basis for an alliance.

For its Green Deal framework, however, Glasgow City Region initially relied on competitive tendering as the central instrument for retrofitting houses. This limited the technical standards for energy performance and provided weak incentives to invest in the necessary skills. This policy framework extended Glasgow's long-time neoliberal policy commitment to new market competition as essential for attracting investment.

In Glasgow's political divergence, a public-good approach has confronted a neoliberal techno-market fix. The labour movement's alternative agenda indicates the necessary political struggle for parallel institutional and sociotechnical change, with relevance to many other localities. As this conflict illustrates, technical criteria always have a sociotechnical character, favouring some social arrangements rather than others.

Rival decarbonization agendas have expressed different sociotechnical imaginaries of the future. For the two levels analysed here:

- Nation-wide decarbonization: In one agenda, a capital-intensive decarbonization technofix justifies indefinitely perpetuating fossil fuels, subordinating trade unions to a cross-class partnership and blurring responsibility for high-carbon continuity. In the rival agenda, fossil fuels would be replaced by renewable energy through various means including workers' cooperatives and public companies.
- House-retrofitting programmes: Within a neoliberal technofix agenda, competitive tendering constrains insulation standards and future improvements. In the rival agenda, a higher energy performance (and thus better heat conservation) would emerge from flexible quality criteria, in engagement with worker–community partnerships involving grassroots innovation; this features a a cooperative eco-localization perspective.

Similar divergences will arise around any low-carbon transition agenda, regardless of its name, for example, Green New Deal, Green Industrial Revolution or Just Transition. International trade unions have implicitly taken a Social Dialogue perspective; they 'endorse the main premises and perpetuate the main approach of the liberal business establishment, of UN agencies like UNEP, of mainstream, "big green" NGOs, and of market-focused think tanks', often within a Green Growth agenda; thus they remain 'captive to a very narrow and de-mobilizing interpretation of Just Transition'. By contrast, a Social Power perspective seeks 'to push trade unionism into a more consciously radical and hopeful space', as advocated by Trade Unions for Energy Democracy (TUED, 2018: 3–4).

Those rival perspectives underlie trade-union divergences over decarbonization pathways. Allied with the fossil fuel industry, many trade unions maintain a commitment to CCS fix, despite its doubtful prospects (TUED, 2018: 3–4). For transformational alternatives, trade unions will need a Social Power approach rooted in civil society initiatives.

Comprehensive decarbonization along socially just lines will depend on political struggles on many levels: to build counter-publics contesting the dominant regime, mobilize frontline communities, build worker–community partnerships, disrupt the hegemonic cross-class partnership for fossil fuels, promote socially just alternatives involving grassroots innovation, demand the necessary resources to implement them and so build collective self-confidence for creating a different future. This process would transform the supporters themselves. Such drivers could push decarbonization agendas towards system change.

7

Conclusion: What Social Agency for System Change?

Introduction

Starting from the popular slogan, 'System Change Not Climate Change', this book has explored its practical meanings. The slogan has targeted high-carbon systems which cause climate change, other environmental harms, resource plunder and social injustices, along with policies which perpetuate them. As the focus here, the system includes the elite's climate fixes for system continuity, their basis in market-driven instruments, and their popular appeal or acceptance for minimizing any societal disruption.

At the same time, public protest and controversy over such fixes has been a political opportunity to advance proposals for replacing harmful production systems with environmentally sustainable, low-carbon, socially equitable ones. But an effective social agency to implement such proposals has been generally elusive. Hence they have remained as technical-administrative blueprints, or appeals to hypothetical planners, or mere supplements to dominant high-carbon systems, especially as fossil fuels gain new investment and total energy usage rises.

An effective social agency would need to come from political forces far broader than those demanding climate action (see Figure 7.1). To explore the broader potential, this book has analysed public controversies where climate fixes entail many other issues. Chapter 1 posed some generic questions:

- How do policy frameworks promote market-type incentives and competition, supposedly in order to generate technological fixes for environmental problems (especially climate change)?
- How do those fixes encourage a passive public to accept or await them, meanwhile continuing harmful production systems?
- How do opponents contest those fixes, stimulate public controversy and so open up different societal futures?

Figure 7.1: XR Trade Unionists (2022)

Note: This network has sought common solutions to overcome intergenerational injustice, racial injustice and environmental destruction. The quote comes from Shelley's famous poem, 'The Masque of Anarchy'.

Source: XR Trade Unionists (2022) Extinction Rebellion Trade Unionists, https://actionnetw ork.org/groups/xr-tu-extinction-rebellion-trade-unionists)

- How do such opponents attempt to build a social agency for alternative solutions?
- In those ways, how do claims for solutions promote divergent societal futures, serving either system change or continuity?

A further overarching question is: How can cross-case comparisons inform strategies for an effective social agency towards system change? Juxtaposing the case studies, here are some generic answers.

Technofix controversy as a political opportunity

For at least half a century, policy elites have regularly promoted technological fixes for various environmental problems and more recently for climate change. Meanwhile they have disparaged or marginalized other available solutions which would be socially and environmentally beneficial. According to some critics, technofixes substitute technological change for the societal change necessary to solve environmental problems.

More subtly, however, the technical/social distinction can be deceptive. Any technical solution structures or favours particular social relations, for example, vertical command-and-control, or market competition or horizontal cooperation. Each corresponds with a future societal vision. So any proposal should be understood and scrutinized as sociotechnical.

In that regard, the case studies here are mainly techno-market fixes, whereby market-type instruments are meant to stimulate technoscientific solutions for environmental problems. Techno-market fixes have sometimes provoked controversy, opening up societal futures different than the fix. The transversal analysis will identify some roles along three lines:

1. How techno-market fixes have made claims for solutions, while blurring responsibility for outcomes.
2. How counter-publics have challenged those fixes, provoking public controversy.
3. How such opponents have promoted different societal futures than the fix does.

In each case study (and more generally), those three roles have been interactive. Techno-market fixes have appropriated key terms from alternatives, especially 'sustainable', 'low-carbon' or 'green' production, thus making the terms more contentious. Meanwhile alternative solutions have helped to oppose and undermine fixes, sometimes appropriating their key terms such as 'climate-smart agriculture', 'knowledge-based bioeconomy' or 'resource recovery' (Chapters 3, 4 and 5, respectively). Nevertheless the three-part structure in this chapter helps to highlight similar patterns across diverse cases (see the summary and comparisons in Table 7.1). Here the patterns recapitulate analytical concepts from Chapter 2.

How techno-market fixes have made claims for solutions and blurred responsibility

Political-economic elites have made recurrent promises for techno-innovations that will fix environmental problems. As parodied by critics, 'The deterioration of the environment produced by technology is a technological problem for which technology has found, is finding, and will continue to find solutions' (Huesemann and Huesemann, 2011: 77). This well describes ecomodernist policy frameworks, which have responded to systemic harms by seeking or claiming eco-efficient technofixes, most recently for climate change. Examples here include agribiotech crops to avoid harms from high-carbon monocultures, advanced biofuels to avoid harms from conventional biofuels as well as from fossil fuels, ATTs to avoid harms from incinerators, and CCS to avoid harms from fossil fuels.

Across those diverse cases, their ecomodernist frameworks express specific economic imaginaries. These have constructed the nation as a competitive economic space, for example, as means towards intellectual property, technology export and advantage over foreign rivals (cf Jessop, 2005). A rhetorical imperative has been economic competitiveness as both means and ends. Techno-market frameworks invoke this economic imaginary for building political alliances, mobilizing resources and justifying market-type policy incentives, which characterize neoliberal environmentalism (cf Levy and Spicer, 2013; see again Chapter 2).

Techno-market fixes both serve and disguise a political agenda, namely: State roles emphasize financial instruments which empower large investors and companies promising eco-efficient technologies, while relegating any responsibility for failures or difficulties to anonymous market forces. Among many potential fixes, priority has been given to techno-innovations suiting market-type instruments – for example, by competing for state subsidy, carbon credits, product markets, intellectual property, and so on – rather than innovations dependent on direct state investment and responsibility. These frameworks depoliticize societal choices, pre-empt alternative futures, marginalize civil society groups, and thus undermine democratic accountability. Both the EU and the UN Climate Convention have played these roles, despite their claims to global environmental leadership (see again Chapter 1).

As in earlier historical periods, calling for technoscientific innovation has been 'a common way of avoiding change' (Edgerton, 2006), or at least any change inconvenient for dominant economic interests. Technofixes have displaced environmental burdens through several means, especially their

Table 7.1: Controversy over climate fixes opening up alternatives

Case-study chapters	Dominant problem-diagnosis and solutions: shifting in response to protest	Counter-publics' problem-diagnosis of the fix	Oppositional alternatives
Techno-market fixes in general	Resource-inefficient technology must be remedied by eco-efficient innovation, stimulated by market-type incentives.	Environmentally destructive, socially unjust production systems generate false promises of future techno-solutions.	Alternatives are resource-light, low-carbon, socially just systems enhancing collective capacities.
Chapter 3: Agribiotech: EU policy and controversy	GM crops with precisely targeted genes will make agriculture more eco-efficient, farmers less dependent on agrichemicals and agri-industry more competitive (mid-1990s). Climate-smart crops will reduce GHG emissions and better withstand environmental stresses (2010s).	Agri-industrial efficiency promotes monocultures dependent on intensive external inputs and degrade natural resources in order to supply global commodity markets. Global market competition drives dependence on such inputs and resource degradation.	Agroecological methods for 'cooling the planet and feeding the people'. Food sovereignty providing democratic accountability and localizing production-consumption circuits. Climate resilient agriculture based on biodiverse agroecosystems.
Chapter 4: Biofuels: EU mandate and controversy	2009 RED will guarantee that biofuels increasingly substitute for fossil fuels in transport and so reduce its GHG emissions. This reduction justifies minimal improvements in fuel-efficiency standards (2010s). Initial market for conventional biofuels and extra incentives will stimulate 2G biofuels beyond edible crops, thus avoiding earlier disadvantages (2010s).	EU policy stimulates greater transport burdens and incentivizes private motor vehicles with liquid fuels and lax fuel-efficiency standards. Market incentive for agrofuels 'drives to destruction', reinforces the internal combustion engine and creates a 'carbon debt time-bomb'. They generate socio-environmentally destructive practices and land grabs in global South.	Transport systems: Better public transport reducing dependence on private transport. More stringent fuel-efficiency standards. Electric vehicles as replacements for internal combustion engine. North–South climate justice: Food sovereignty for localizing production-consumption circuits, avoiding high-carbon inputs, and resisting global market competition.

Table 7.1: Controversy over climate fixes opening up alternatives (continued)

Case-study chapters	Dominant problem-diagnosis and solutions: shifting in response to protest	Counter-publics' problem-diagnosis of the fix	Oppositional alternatives
Chapter 5: Waste-conversion: UK policy and controversy	MSW incineration plants make improvements, bring waste up the hierarchy and reduce GHG emissions relative to landfill (2000s). ATTs will overcome limitations of incinerators, help localize waste management and provide renewable substitutes for fossil fuels (2010s). MSW incineration plants are a transitional technology for moves towards a circular economy and net-zero carbon.	Linear economy of 'use and dispose' structurally wastes resources. Incinerators generate pressures for waste supplies 'feeding the beast'. Policy-financial incentives prioritize resource recovery, especially electricity supplying centralized systems. False promises of future solutions perpetuate demand for more waste feedstock and its wasteful use.	Circular economy: Redesign production systems to reduce waste generation, to make waste more easily re-usable and address socio-economic inequities. MRFs should help avoid new incinerators, not simply accompany them. MRFs can stimulate recycling and bring community benefit via materials reuse and renewable energy.
Chapter 6: Green Growth versus GND agendas. Retrofitting: Techno-market fix versus public good	CCS will reconcile economic growth with decarbonization of fossil fuels, as a basis for skilled trade-union jobs and technology export. Within the Glasgow Green Deal, competitive tendering for retrofit contracts will bring houses up to EPC band C.	As a false solution, CCS promises serve to continue fossil fuels, to reinforce the industry's power and to delay a low-carbon transition. For retrofitting houses, competitive tendering limits energy-efficiency standards, fragments the necessary expertise, marginalizes labour movement contributions and favours large foreign companies for contracts.	Fossil fuels should be replaced by renewable energy on a socially just and politically accountable basis. Retrofit programmes should ensure the public good through a municipal company, partnerships with the labour movement, local manufacturing capability and higher standards.

transformation, relocation and/or time-delay (Lecain, 2004), as in some case studies here. Techno-optimistic expectations have provided an alibi to continue carbon-intensive systems while seeking or awaiting market-driven low-carbon innovation, sometimes providing substantial state funds for their development. Techno-innovation promises have evaded a low-carbon, socially just transition, much less a socially just process. Thus such fixes pre-empt or delay beneficial alternatives. This overview summarizes such patterns across the case-study chapters.

The EU agribiotech fix (Chapter 3)

Since the 1960s, intensive pesticide usage has caused significant harm to flora, fauna and human health. Widespread protest demanded restrictions or bans on many pesticides. In the 1980s some agrichemical companies developed genomics techniques for a new agribiotech (or Life Sciences) industry, promising novel products that would reduce farmers' dependence on agrichemicals and increase yields. According to its promises, agri-innovations would substitute knowledge of biomaterials for chemicals, target pests precisely, provide eco-efficient resource usage and so conserve natural resources. Conventional crops were portrayed as genetically deficient for self-protection from environmental threats, thus needing biotechnological improvements. Agribiotech companies deployed such techno-optimistic promises as a political imperative for state R&D funds and greater market pressures favouring capital-intensive agri-inputs for higher yields. This Life Sciences sociotechnical imaginary underlay the EU's state–industry partnership, aiming to make agri-industry more globally competitive.

From the 1990s onwards the EU elaborated this neoliberal agenda through several institutional changes. These have featured genomics-level agri-research priorities and broader intellectual property rights, facilitated by a draft EC Directive on 'biotechnological innovations'. Together these were favourable for privatizing genetic resources and knowledge, thus provoking disputes over 'biopiracy'. After a decade-long controversy, the Directive was enacted in 1998; then some EU member states sought to block its implementation. This set the stage for a broader controversy over agribiotech in the late 1990s.

The EU biofuels fix (Chapter 4)

For several decades, environmentalist groups have highlighted a major threat from the EU's rising transports, fossil fuel usage and thus GHG emissions. This rise has been driven partly by the EU's 1990s infrastructural agenda to accelerate EU-wide transport flows, intensify market competition, increase productive efficiency and thus enhance international competitiveness. This policy driver of greater GHG emissions was evaded by the EU agenda for mandatory biofuel targets, as legislated in the 2009 Renewable Energy Directive. The EU mandate had several institutional roles, namely: to make the EU a buyers' market for competing biomass suppliers, to subsidize EU farmers producing such feedstock, to finance R&D for future biofuels which could gain intellectual property, to strengthen the 'investment climate' for companies creating bio-innovations, and to normalize the EU's rising transport flows. Such political-economic roles comprised an economic imaginary of an investment climate, which took priority over climate objectives.

The RED was officially meant to reduce GHG emissions from transport fuel by mandating a percentage from renewable energy. Promises for biofuels as low-carbon sources provided an alibi to expand transport flows indefinitely, and as well as to postpone tighter standards for fuel efficiency. As a techno-market fix, the mandatory biofuel targets created or expanded a market for biomass feedstocks, incentivized high-carbon methods, intensified global resource extraction, and aggravated land-use rivalry between locally consumed food versus fuel (along with animal feed). By inducing indirect land-use change, moreover, these harms were extended beyond the narrow scope of EU sustainability criteria. Meanwhile those criteria served to greenwash the EU's global resource plunder.

Biofuel targets were meant to stimulate investment in the EU's KBBE. Alongside the economic imaginary of a more competitive Europe, this sociotechnical imaginary anticipated eco-efficient technoscientific innovation for more lucratively using bio-resources. In particular, crop cultivation on 'marginal land' would provide (non-edible) cellulosic biomass for 'advanced' biofuels. Such future arrangements were meant to avoid harm from conventional biofuels. Yet any such technological advance remained elusive for at least a decade after the 2009 Directive. Meanwhile the promise of future 'sustainable biofuels' reinforced dependence on the internal combustion engine and thus system continuity.

The UK's waste-conversion fix (Chapter 5)

The UK has had an institutional commitment to market competition as a means to generate techno-innovations that could reconcile diverse objectives, especially at the nexus of waste management and renewable energy. From 2006 onwards the UK's Waste Infrastructure Delivery Programme for managing MSW was meant to bring many environmental and economic benefits. The programme was driven by financial incentives to reduce landfill (as mere disposal) and improve resource recovery through renewable energy production, whose substitution could reduce GHG emissions.

This programme mainly subsidized contracts for large mass-burn incinerators, provoking widespread protest. These incinerators relocated environmental burdens by increasing long-distance waste transport, as well as by transforming MSW (including plastic) into carbon emissions. Moreover, such plants were generating demand for more waste 'to feed the beast' and so arguably undermined efforts to increase recycling.

As a techno-market fix for waste disposal and related environmental burdens, extra financial incentives were meant to stimulate novel Advanced Thermal Treatments (ATTs); these would turn MSW into high-value products, thus overcoming the limitations of incinerators for resource recovery. Techno-optimistic assumptions further envisaged small-scale technology helping to localize responsibility and decentralize material flows through renewable substitutes for fossil fuels. Yet ATTs extended the problems of conventional incinerators, encountered operational difficulties with heterogeneous waste feedstock, imposed environmental nuisances on low-income areas, and so still provoked opposition. Market-type policy instruments incentivized conversion processes with dubious environmental advantages over landfill, especially electricity production with long-distance material flows. The policy framework relegated waste-management decisions to market competition, thus blurring responsibility for various technical failures and resource-burdensome outcomes.

The Carbon Capture and Storage (CCS)-fossil fuels fix (Chapter 6)

Protest against climate change has raised political demands to decarbonize the economy by substituting renewable sources for fossil fuels. A joint state–industry response has been to promise future technologies to decarbonize fossil fuels. This has protected the enormous sunk investment in coal, gas and oil. In particular, the Green Growth agenda has promoted CCS, seeking to reconcile economic growth with decarbonization and environmental protection. The CCS fix promises to transform and relocate carbon to storage sites, dependent on optimistic assumptions about their renewable energy inputs and long-term stability; this provides a basis for displacing environmental problems across space and time. This agenda expresses an economic imaginary of the nation as a unitary competitive space, whereby technoscientific advance will provide a global competitive advantage for technology exports.

That economic imaginary has complemented a sociotechnical imaginary whereby decarbonization technologies (especially CCS) will enable fossil fuels to continue, as a basis to provide secure, skilled employment. Trade unions in high-carbon sectors have demanded a commitment to CCS as a basis to support low-carbon transition agendas such as a Green New Deal or Just Transition. Promises to reconcile all those aims (environmental, economic and social) have softened societal conflicts over a potentially disruptive low-carbon transition; this illustrates a popular ambivalence towards climate fixes.

In practice CCS has provided an alibi for continuing fossil fuels, reinforcing the industry's political-economic power, and blurring responsibility for high-carbon outcomes. The dominant sociotechnical imaginary underpins a cross-class partnership justifying system continuity. The political task remains to undermine that partnership, as a prerequisite for labour movement leadership in decarbonization agendas.

How counter-publics challenged those fixes and provoked public controversy

Techno-market fixes for environmental problems have encountered criticism, often generating public controversy. The case studies here had some general patterns: The techno-market fix was promoted as simply a technical solution to remedy environmental problems (for example, resource degradation, resource burdens and/or carbon emissions), but in practice reinforced their causes in production systems. Critics generated public controversy about the putative environmental benefits, which were contradicted by current or plausible harms.

Each controversy entailed various issues such as health, environment, climate, resource degradation, socio-economic inequities, North–South exploitation, and so on. As a feature of social movements, 'frame-bridging' aligns 'two or more ideologically congruent but structurally unconnected frames regarding a particular issue or problem' (Snow et al, 1986: 467). In the cases here, frame alignments deployed pejorative metaphors and analogies. For example: Agribiotech 'efficiency' claims were associated with the mad cow pandemic as agri-industrial hazards. 'GM contamination' was associated with various threats to safety and democratic accountability. 'Agrofuels' were associated with chemical-intensive high-carbon agriculture and a 'carbon emissions-debt time-bomb' from land clearances for cultivation. Incinerators were associated with ominous pressures 'to feed the beast'; ATTs were called 'incinerators in disguise'; their waste burdens were attributed to the linear economy of 'use and dispose'. With those bridging frames, opponents aligned various critical frames, evoked analogous controversies, linked failures or harms of each fix, and undermined expert claims of political neutrality.

In each controversy, elite techno-optimistic expectations were challenged by citizens' groups, which can be understood as mobilized counter-publics. They integrated diverse knowledge sets, criticized dubious assumptions, identified 'undone science' and sometimes filled knowledge gaps (cf Hess, 2007, 2016; Frickel et al, 2010). Such groups contested the official expertise as partisan, for example, for exaggerating environmental benefits, accepting lax standards, concealing socio-economic inequities and privileging dominant economic interests. Counter-expertise undermined official assumptions in several ways, for example, by highlighting evidence of malfunction or harm, and attributing this to systemic causes in market drivers.

State authorities generally have framed the issues as risk or sustainability in a narrow sense, whereby a novel technology or product may directly cause biophysical effects. To evaluate such potential effects, regulatory expertise has implicitly depended on normative assumptions as regards what potential effects would be ir/relevant, un/acceptable or better/worse than a putative baseline. With such normative assumptions, regulatory procedures initially made favourable judgements.

In response, counter-publics have gone beyond or contested those normative assumptions through plausible effects. For example: for GM crops, agronomic changes in herbicide usage could harm farmland biodiversity; for biofuel feedstocks, global commodity markets could stimulate crop displacement, land-use change and greater GHG emissions; and for some waste incinerators, the thermal treatment may provide only resource disposal and increase GHG emissions. Public controversy was extended to regulatory expertise as politically partisan.

To accommodate such criticisms, the regulatory procedure has sometimes adopted more stringent criteria, which posed greater evidential burdens for the fix. Neoliberal regimes have sometimes flexibly responded with such neo-regulation as means to structure, adjust and legitimize new markets (Busch, 2010: 334). Moreover, deploying more stringent regulation, counter-publics have found opportunities to undermine or constrain the technofix; they have also highlighted its systemic drivers and institutional commitments. These oppositional strategies sometimes emerged more clearly from researchers' discussion with activists through a Participatory Action Research process (see penultimate section). Let us look more specifically at each case.

The EU agribiotech fix (Chapter 3)

The EU's agribiotech agenda was turned into a controversy over predictable and unknown risks. Many critics emphasized genetic novelty as a potential source of future hazards. Critics also diagnosed agri-industrial efficiency as a threat, drawing analogies to the 'mad cow' pandemic; thus they gave ominous meanings to the industry's eco-efficiency claims for agribiotech. As some early critics also warned, GM crops were designed for monocultures dependent on intensive external inputs which degrade natural resources in order to supply global commodity markets. As a basis for intellectual property, agribiotech further undermined farmers' knowledge and the seeds-knowledge commons.

Through mass protest from the late 1990s onwards, GM products were popularly stigmatized as a threat to the environment, human health and sustainable development. In response, the EU regulatory procedure more broadly evaluated how changes in agronomic practices (for example, herbicide usage with herbicide-tolerant crops) could cause environmental harm of more diverse kinds. Companies faced greater regulatory burdens and were denied regulatory approval of some GM crops.

Moreover, critics popularized an evocative metaphor, 'GM contamination'. This linked several threats such as ubiquitous pollen, uncontrollable hazards, genetic uniformity, lost income of organic farmers, and democratically unaccountable decisions. Along with agri-industrial 'efficiency', the pejorative phrase 'GM contamination' served a frame-bridging role, aligning otherwise disparate issues.

'The market' acquired contradictory meanings: The EU's neoliberal policy framework for agribiotech had been officially justified as necessary for adapting to competitive-market forces. As a different meaning, food markets became vulnerable to consumer protest against GM products. Each European supermarket chain had based its reputation on 'own-brand' products and eventually declared that they would exclude GM ingredients. Meanwhile more local authorities declared 'GMO-free zones', extending to entire 'GMO-Free Regions', seeking especially to avoid 'GM contamination' of non-GM crops. Using such opportunities, a broad opposition network blocked an EU-wide market for GM products and constrained field trials.

EU biofuel fix (Chapter 4)

To justify the EU's biofuels mandate, its advocates had portrayed biomass feedstock as lower-carbon than fossil fuels. In response, civil society groups blamed 'agrofuels' as an agri-industrial source of several threats: indirect land-use changes (ILUC), rivalry between 'food versus fuel', chemical-intensive methods and thus higher carbon emissions. Thus the pejorative term 'agrofuels' served as frame-bridging to align otherwise disparate frames.

Critics warned against global commodity markets stimulating ILUC as a systemic threat causing multiple harms including GHG emissions. Early on, North-South development networks cited ILUC effects as an extra reason to oppose the EU biofuels mandate. Even before the 2009 RED was enacted, large environmental NGOs were jointly opposing the EU plan 'to massively expand the use of biofuels', that is, an obligatory rise in the percentage in transport fuel. As a common criticism, false carbon accounting downplayed or disguised significant emissions; weak sustainability criteria allowed feedstocks which caused the greatest harm.

In opposing the EU mandate, critics encountered several obstacles. The agri-industrial farmers' lobby was strongly allied with the capital-intensive industries promoting biofuel expansion. Although biofuel criticisms were accepted by many politicians, some anticipated that opposing or limiting the biofuels mandate could jeopardize the Directive's overall support for renewable energy. By contrast with the earlier dispute over 'food versus fuel', the ILUC issues seemed too complex for large NGOs to mobilize their support base against the EU biofuel mandate.

ILUC issues were channelled into the EU's expert procedures to evaluate the various feedstock crops, generating expert disputes over global trade models. Meanwhile the mandate's proponents defended the broad feedstock eligibility as necessary to maintain 'an investment climate' for the KBBE, as symbolized by biofuels. The 2009 EU mandate continued to incentivize the most harmful feedstocks until at least a decade later. By then, the carbon intensity of EU transport fuel was barely lower than before the 2009 mandate. Nevertheless the techno-market framework had played its implicit political-economic role for system continuity.

The UK's waste-incineration fix (Chapter 5)

The UK Energy-from-Waste (EfW) incinerator programme provoked rising protest. Local opponents highlighted health and environmental harms, which were generally marginalized and dismissed by expert-regulatory procedures. From a systemic perspective, opponents denounced incinerator expansion for pressures 'feeding the beast', perpetuating demand for waste, undermining waste-recycling, contradicting the waste hierarchy and thus wasting resources. The incineration industry had internal disagreements about environmental advantages of various thermal options, so opponents highlighted those disagreements as a basis to discredit all thermal treatments. Some proposals for new plants underwent more stringent environmental criteria, for example, greater scrutiny of claims for resource recovery rather than mere disposal, and comparisons with lower GHG emissions as the appropriate normative baseline.

As climate change became a more salient issue, the anti-incineration campaign made greater efforts to undermine the industry's claims for GHG savings. Local campaigns were being joined by Extinction Rebellion Zero Waste (XRZW), which gained significant organizational signatories to an Open Letter. This warned that the great expansion of EfW incineration plants would increase carbon emissions. Therefore the UK government should transform its waste and resource sector 'to accelerate the transition towards a genuine, zero-waste circular economy', rather than accept more incinerators. By 2021 local waste authorities had been deterred from approving more than 70 new incinerators. These conflicts deepened the incineration controversy, towards a focus on resource conservation and reuse in production systems.

Carbon Capture and Storage (CCS) divergences around Green New Deals (Chapter 6)

Elite agendas have promoted CCS as a supposed means to decarbonize fossil fuels, with support from trade unions in those sectors. Critics have warned against this solution as elusive, false or even environmentally harmful. Amidst many technical failures, CCS has most succeeded in installations extracting oil, thus undermining climate objectives. Such warnings have been raised by a global network of labour-movement groups, especially public-sector trade unions (TUED, 2015, 2018). Thus the labour movement has a deep divergence over means towards a low-carbon transition.

This divergence arose within trade-union confederations in the UK and US. Likewise in both countries when GND proponents sought support for their strong decarbonization agendas from a major political party. These national efforts had different outcomes: US trade unions in high-carbon sectors criticized the US Justice Democrats' GND for omitting CCS. By contrast, their UK counterparts successfully incorporated CCS into the Labour Party agenda for a GND, thus undermining its decarbonization aims.

CCS-fossil fuel fixes became central in the 2019 GND agenda of the UK Labour Party and likewise in the 2021 Climate Jobs plan of the Biden Administration. Both complement capitalist Green Growth agendas for supposedly reconciling economic growth with decarbonization. They maintain capitalist leadership and reinforce system continuity, backed by a cross-class partnership of high-carbon energy producers and their sector's trade unions. All this marginalizes various public-good alternatives such as worker–community cooperatives, environmental amenities and renewable energy truly replacing high-carbon sources in overall energy usage.

How opponents have promoted alternative futures

In each case study here, opponents denounced the technofix for perpetuating systemic causes of environmental problems. They counterposed alternative solutions, often expressing an eco-localization imaginary. Such agendas would localize production-consumption circuits and responsibility for them through collective learning processes (cf North, 2010: 588; see also Feola and Jaworska, 2019).

Some cases feature grassroots innovation, a socially inclusive process involving users in the design or even users becoming the innovators (Smith et al, 2014, 2016; Smith and Stirling, 2016). Often the innovation process has built a commons in knowledge and resources, while resisting their private appropriation (cf de Angelis, 2003; also 2017). Each design is sociotechnical, linking technical, institutional and societal change in distinctive ways. Drawing on those concepts, let us survey the case studies in turn.

Agroecology versus agri-industrial systems (Chapter 3)

Shortly after the turn of the century, agribiotech opponents counterposed biodiverse 'quality' agricultures. The Assembly of European Regions declared GMO-Free Regions, which promoted such alternatives. By a decade later, farmer–civil society alliances were promoting agroecological production methods. These depend on farmers' grassroots innovation for both using and conserving biodiversity as a resource commons. Support networks have devised short supply chains bringing producers closer to consumers, along lines expressing eco-localization imaginaries.

Europe-wide farmer–civil society networks have promoted agroecology support measures through adaptations or changes in the EU's CAP, though they have made only modest gains. As a main reason, dominant political parties have supported the farmer–agribusiness lobby, defending subsidy criteria which perpetuate external high-carbon inputs to raise productivity. Again this illustrates rival sociotechnical imaginaries of a future society.

When climate change became a more salient issue, GHG emissions from agriculture were turned into a general focus of debate. In 2014–2015 a joint state–industry agenda promoted 'climate-smart agriculture' whereby agricultural land would become a more effective carbon sink through no-till methods, facilitated by external inputs including herbicide-tolerant GM crops. Such land could qualify for tradeable carbon credits and expand a market for them, thus providing a techno-market climate fix. A broad network of civil society groups denounced this as 'corporate-smart greenwash'. They turned the fix into a controversy over the intensive monoculture practices that degrade natural resources and increase GHG emissions. The state–corporate agenda for tradable carbon credits lost impetus for several reasons: critics undermined its credibility, a low carbon price weakened any financial advantage, and agricultural land was anyway shifting from a carbon sink to a carbon source. Agribiotech opponents promoted agroecological methods for 'cooling the planet and feeding the people'; this aligned their various critical frames, thus strengthening a coherent opposition and alternative.

Alternatives to agrofuels (Chapter 4)

Since well before the 2009 EC RED, diverse campaign groups opposed the EU biofuels mandate for causing multiple harms. They counterposed various alternatives, for example, stricter fuel-efficiency standards, a shift to electric vehicles (with renewable energy), better public transport, agroecology for food sovereignty, biodiversity conservation, and so on. Although those alternatives had some synergies, campaign groups had disparate emphases and issue-frames. These would have posed difficulties for a joint campaign, especially common demands for an alternative future.

Nevertheless, their alternative agendas reinforced and expanded the North–South networks that had originally initiated wider opposition to the EU biofuels mandate. Their efforts alerted European civil society groups about elite claims for 'sustainable' resource usage in the global South, by contrast with the everyday realities of global market forces driving plunder. Their warnings built a broader support base for later interventions such as the 2014–2015 dispute over 'climate-smart agriculture' (see again Chapter 3).

Circular economy versus incineration (Chapter 5)

In opposing new waste-incineration plants, the UKWIN network pejoratively associated these with the linear economy of 'make anew, use and dispose'; this systematically wastes resources and creates hazards. They counterposed a circular economy agenda of 'reduce, repair, reuse, recycle' waste. This had several features: locally-based production-consumption circuits, production designed to minimize waste-generation, more effective waste collection, greater waste recycling, renewable energy production, and so on. Some benefits were meant to be integrated through Materials Recovery Facilities (MRFs). These were anyway being promoted to greenwash incinerators; by contrast, anti-incineration campaigns demanded that MRFs should substitute for any new incinerator capacity. In these ways, circular versus linear economy served as frame-bridging, aligning diverse frames of anti-incineration critics.

Although the UK policy framework included notions of a circular economy, this could remain as merely technical-administrative proposals, or as a drive for greater resource recovery from waste, or as a business agenda to increase private profit. By contrast, anti-incineration campaigns advocated a circular economy as a socially just systemic change. They advocated means for production systems to reduce waste at source and for residual waste management to provide societally equitable benefits, especially to address fuel poverty. Some local anti-incineration groups were building or joining political alliances for such publicly accountable alternatives. These have drawn on eco-localization imaginaries to prioritize localized, socially equitable uses of waste. Such proposals could be held publicly accountable, even popularly shaped, thus gaining a transformative potential.

Green New Deal local agendas: retrofitting houses (Chapter 6)

Labour movement groups have been initiating GND local agendas, which seek to replace high-carbon production systems with low-carbon, socially equitable alternatives. For heating houses, the UK has depended mainly on natural gas, so its replacement became a major focus of government policy and of rival solutions. National trade-union bodies endorsed the gas industry's CCS-hydrogen agenda for supposedly decarbonizing natural gas, as initially piloted in Leeds. Nevertheless in 2020 Leeds Trades Council decided to oppose this agenda and undermine its cross-class partnership. For a local GND, it advocated a worker–community partnership for retrofitting houses with insulation and heat pumps, whereby the local authority would in-source expertise and responsibility. This proposal was widely circulated around labour movement networks.

As a major opportunity, in 2021 the Scottish government gave Glasgow City Region substantial funds to begin a retrofit programme. Although this was labelled a GND, the local authority extended its long-time neoliberal policy framework: it relied mainly on competitive tendering for retrofit contracts. This techno-market fix favoured a minimal standard of energy performance and deterred knowledge exchange among prospective contractors towards achieving better-quality standards. As a different version of a Green New Deal, an alliance between the Scottish labour movement and civil society groups has advocated a policy shift towards flexible retrofit designs for greater energy performance, linked with worker training for such standards. This arrangement would encourage grassroots innovation and a cooperative knowledge commons for technical design with local resources, resonating with eco-localization imaginaries. Those rival retrofit agendas appropriated the GND banner for contrary aims: the local authority agenda for neoliberal system continuity, versus the civil society agenda for system change.

Towards an effective social agency: collective capacities for system change

As in the case studies here, political protest has recurrently created public controversy over high-carbon systems and deceptive fixes for them. To transform or replace those systems, critics have counterposed low-carbon, socially just alternatives. What means are necessary to advance and implement them, especially as steps towards system change?

Despite great efforts, it has been difficult to create an effective social agency (see again Chapter 1). This would depend on multi-stakeholder convergences, collective capacities and adequate resources to transform or replace harmful high-carbon production systems, especially dependence on high-carbon imports from the global South. How could this effort prevail against strong political-economic forces of system continuity? As a reference point, let us return to a political initiative from Chapter 1.

Under the dual banner of a GND and Just Transition, an EU-wide political alliance has advocated a *Green New Deal for Europe* (GNDE). This has sought to transform the EU's neoliberal regime through both confrontation and proposition. The GNDE advocates 'a logic of confrontation pitting Europe's communities against the EU institutions that seem unwilling to see the climate and environmental crisis through the lens of their lived realities'. The crisis has been aggravating the harms from fiscal austerity regimes, as enforced by the EU. 'Communities that have already been pushed to the margins of the economy are often at the frontline of the climate and environmental crises.' Moreover, the EU outsources GHG emissions and other environmental harms to the global South, as a basis for deceptive claims about Europe's decarbonization. Europe has been 'exporting unsustainable practices beyond its borders' and along its supply chains, thus extending Europe's colonial legacies through today's green colonialism (GNDE, 2019: 21, 77; see also excerpts in Chapter 1).

That critical perspective resonates with many social movements confronting the neoliberal, neocolonial system that worsens climate change. Activists with such critical perspectives have participated in XR. It has targeted the high-carbon economy and complicit institutions; such actions have highlighted systemic causes of climate change, while overcoming many people's fatalistic passivity. XR offshoots include XRZW, promoting low-carbon alternatives to new incinerators, as well as Insulate Britain, which has raised support for low-carbon retrofits (see Chapters 5 and 6). Many activists have questioned the political system as an adequate basis to implement climate solutions and have discussed what social agency could do so.

XR has demanded a shift from the 'toxic system' that causes climate change, yet the XR leadership has been silent on its key driver in profit-driven economic growth (Stuart, 2022). Likewise evading the necessary social agency for system change, the leadership has made appeals to 'Go Beyond Politics'. Namely, citizens' assemblies would use deliberative democracy, devise recommendations and demand their adoption by state authorities; this strategy would 'reclaim power from the bottom up'. Such change could happen 'if the pressure of a citizens' assembly is larger than the pressure of a government's supporters and funders' (XR, 2021). This scenario implausibly assumes that a citizens' assembly per se could push the state beyond its neoliberal form and its persistent role in system continuity (as illustrated in the case studies here).

On the contrary, a socially just low-carbon transition would depend on a power struggle for institutional change, which has been hardly identified, much less advocated. To fill the gap, the GNDE proposed new EU structures for horizontal cooperation that would share best practices and expand administrative capacities for such alternatives. Rather than await such institutional changes, moreover, grassroots organizations and communities should mobilize 'to make the vision a reality' (GNDE, 2019: 7, 9).

Yet how? This question can be answered only in practice by a collective social agency testing specific strategies. This includes several tasks: to bridge diverse frames, build political alliances, to confront socially unjust high-carbon systems, to promote alternative futures, to seek the necessary resources and to evaluate outcomes, as a basis to reconsider or refine the strategies. As a contribution here, the case studies had some patterns which were identified through a conceptual framework. A transformative mobilization would need to combine five elements: mobilized counter-publics, frame alignments, eco-localization imaginaries, grassroots innovation and solidaristic commoning for the necessary resources (as elaborated in Chapter 2).

To integrate those elements needs several collective capacities. 'Collective' means that each participant or group brings various strengths that can be combined and strengthened through joint efforts. Collective capacities are necessary for the roles discussed in the following subsections.

Frame alignments towards a common vision

A socially just low-carbon transition has a task to align the issue-frames of diverse groups. These involve tensions and awkward trade-offs, as illustrated by the labour movement's divisions over decarbonization (Chapter 6). Through political debate on societal futures, some framings and alignments may attract greater support, contributing to a common vision.

For many groups, an entry point may be socio-economic or environmental injustices, especially by toxic industry or resource extraction which maintains high-carbon systems. Climate Justice perspectives have stimulated more societal groups to confront those systems and engage with decarbonization agendas. Through multi-issue environmental conflicts, the term 'frontline communities' has gained broader meanings beyond its origin in the global South or near US toxic industries.

Likewise the 1990s demand for 'environmental justice' has become broadened through the slogan 'climate justice', encompassing several groups facing various threats. More generally:

> In the Global North, we have workplace struggles of migrant workers who perform the bulk of care work in homes and hospitals, alongside a growing strike wave led by teachers and nurses. These are joined by community struggles for clean water and clean air, most often led by communities of color, exposing the deliberate, racialized poisoning of the environment by capital. (Bhattacharya, 2019)

For example, incinerator programmes extend colonial legacies to environmental racism at home, thus worsening climate injustice and stimulating more opponents. In the UK a broad network has been opposing new incineration capacity as a multiple threat – on health, environmental, racism and climate grounds. They counterpose resource-recovery facilities, towards a future socially just circular economy.

Collective capacities are necessary for bridging such diverse frames through a comprehensive future vision. Differences in framing environmental problems arise from substantive differences in people's life-experiences and normative standpoints. In exploring those differences, it is necessary to identify how more inclusive framings could inform more effective action. This difficult task can be facilitated by more societally relevant research (Bond, 2018: 179). This leads to the next kind of capacity.

Participatory Action Research for a systemic intervention

For several decades, PAR has structured researchers' collaboration with civil society partners so that they can become more effect agents for societal change (see Chapter 2). PAR helps to explore societal problems and solutions from the partner's standpoint, as a basis to clarify its intervention strategies. In a well-known schema for joint learning, each cycle proceeds from an action plan, to implementation, to observation of outcomes and then reflection towards the next cycle (Heron and Reason, 2006, 2008). These strategies warrant evaluation for their counter-hegemonic aims, emancipatory potential and practical effectiveness (Jordan and Kapoor, 2016).

In this book, the relevant action includes efforts to undermine techno-market fixes, while also counterposing alternatives. Such PAR experiences have informed the case studies. PAR methods valorized the experiential knowledge of societal groups who were harmed or marginalized by techno-market fixes; this 'extended epistemology' helped to contest the hegemonic one (Heron and Reason, 2008). Through such collaborations, civil society partners have deepened a systemic perspective on capital-intensive innovation which supposedly bring eco-efficient solutions for decarbonization or sustainable development. Research partners criticized false solutions for perpetuating unsustainable, high-carbon unjust resource usage. These have been underpinned by regulatory criteria dependent on normative assumptions, contrary to the pretence of science-based political neutrality, as contested by counter-publics.

These understandings have been deepened through knowledge-exchange between professional and practitioner cultures, dependent on bridging conceptual or terminological differences. PAR methods brought an interactional expertise for creating meaningful exchanges between scientists and non-scientists; likewise for exchanges between social scientists and peasants, thus overcoming some barriers (Levidow and Oreszczyn, 2012: 37). Civil society groups thereby broadened their strategies for multi-stakeholder partnerships and policy interventions (relevant to Chapters 3 and 4 here; Hinchliffe et al, 2014).

As one case, the EU's agri-research priorities favoured reductionist capital-intensive approaches, hence marginalizing farmers' knowledge.

Through our PAR process, the civil society partner strengthened knowledge exchange about agroecological methods (among French peasants and with scientists), as a basis to shape research agendas and to gain public support. The PAR methods stimulated multi-partner discussions on institutional obstacles to research cooperation and identified ways to overcome them, so that agroecology could become more effective as an alternative agenda for research and practitioners (for example, Levidow and Neubauer, 2012).

As regards the EU biofuels mandate, the pejorative term 'agrofuels' more readily attracted social movement networks in the global South. In our PAR process, the civil society partner led research clarifying analogies with drivers and multiple harms from wider agri-industrial systems. These analogies strengthened intervention strategies to challenge the EU's beneficent assumptions about 'sustainable biofuels' (Franco et al, 2010) and likewise the EU's investment priority for a future bioeconomy (TNI, 2015).

Research partners thereby strengthened their capacities to identify the drivers of unjust production systems, to challenge techno-market fix agendas and to counterpose alternatives through alliances at several levels. This research project was able to obtain substantial funds for their staff time. But such formally structured, funded projects have been rare.

More generally, researchers have often engaged civil society groups in strategy discussions on systemic drivers of socio-environmental harm. They have jointly analysed policy frameworks which incentivize harmful production systems, critical interventions contesting technofixes, their practical outcomes and alternative futures. These discussions have brought mutual insights about issue-framing, institutional assumptions, alliance-building, political outcomes and lesson-drawing for future conflicts. Those efforts indicate how PAR can contribute to collective capacities and strategies of a social agency towards system change.

Sociotechnical imaginaries for system change

In pervasive disputes over decarbonization pathways, rival agendas have been analysed here as sociotechnical imaginaries. Each envisages or advocates distinctive sociotechnical means towards a feasible, desirable future (see Chapter 2). Dominant imaginaries emphasize market-type incentives for capital-intensive techno-innovation as essential for societal benefits, thus provoking critical scrutiny and sometimes opposition. By contrast, alternatives elaborate their own sociotechnical imaginaries such as eco-localization. These emphasize group learning processes to develop innovation for socially just futures along publicly accountable lines, while counterposing these to the dominant model.

Integral to sociotechnical imaginaries, any 'technical' design or standard should be understood as sociotechnical, favouring some socio-economic arrangements rather than others. At least implicitly, any techno-innovation involves normative commitments to a specific social order. Likewise it involves assumptions about the necessary resources (such as raw materials, energy and labour), which relate to diverse meanings of key terms – such as sustainable, green, smart and low-carbon. These terms warrant scrutiny and debate.

Moreover, a socially just, low-carbon future would need to change the prevalent meanings of economy, work and employment. Necessary changes would re-orient work around 'life-making jobs' rather than endlessly producing commodities for the sake of profit or employment growth (Bhattacharya, 2019). 'Green' would need to mean different society–nature relationships, beyond simply less-polluting processes or greater resource efficiency.

Towards the necessary social agency, a low-carbon, socially just transition would depend on 'a massive, broad-based effort to radically reduce the influence that corporations have over the political process' (Klein, 2019: 90). Likewise it would depend on replacing market competition with solidaristic cooperation. In renewable energy production, for example, state support measures often create or reinforce competitive pressures that fragment knowledge and limit alternatives. So these measures must be replaced with ones that facilitate cooperative learning (TUED, 2020).

For a transformative agenda, we 'will need the skills and expertise of many different kinds of storytellers: artists, psychologists, faith leaders, historians, and more', notes Naomi Klein (2019: 271). These efforts may be informed by concepts such as: eco-localization based on short supply chains, a sufficiency economy to reduce material burdens, grassroots innovation and a commoning process which creates and manages common resources. These can be encompassed by a solidarity economy (Miller, 2010; RIPESS, 2011, 2015; Alexander, 2015; North et al., 2017, 2020). An essential capacity lies in collectively imagining, piloting and evaluating such efforts as a group learning process.

Let us return to the aspiration for 'System Change Not Climate Change'. This slogan originated in the Climate Justice movement and eventually gained a wider appeal. Its perspective has help to diagnose the problem as the political-economic system which perpetuates climate change, other environmental harms and socio-economic injustices, partly through decarbonization fixes.

From a Climate Justice perspective, political protest has confronted such fixes and counterposed low-carbon, socially just alternatives. Across diverse cases and contexts, the analysis here has identified a big picture. Transformative mobilizations need to combine five main elements: mobilized counter-publics, frame alignments, eco-localization, grassroots innovation and solidaristic commoning. Together these can help build an effective social agency for system change.

References

350.org (2021) Lawmakers worldwide unite under new 'Global Alliance for a Green New Deal', *350.org*, https://350.org/lawmakers-worldwide-unite-under-new-global-alliance-for-a-green-new-deal

ABN (2007) *Agrofuels in Africa: The Impacts on Land, Food and Forest*, African Biodiversity Network (ABN), www.gaiafoundation.org/documents/Agro fuelsAfrica_sml_Jul2007.pdf

Action Aid (2012) *Fuel for Thought: Addressing the Social Impacts of EU Biofuel Policies*, Brussels: Action Aid International.

Action Aid Intl (2014) *Corporate-Smart Greenwash: Why We Reject the Global Alliance on Climate-Smart Agriculture*, Johannesburg: Action Aid International, www.iatp.org/files/open%20letter%20against%20GA CSA%20for%20BKM%20summit.pdf

Action Aid UK (2013) *Drive Aid* [film], smarterfuelfuture.org/blog/details/actionaid-drive-aid/

AEEU (2020a) *Agroecology Initiatives in European Countries: Key Findings & Recommendations*, Corbais: Agroecology Europe (AEEU), www.agroecology-europe.org/wp-content/uploads/2020/11/Mapping-report-key-findings-and-recommendations-final-for-circulation.pdf

AEEU (2020b) *Agroecology Initiatives in Europe*, Corbais: Agroecology Europe (AEEU), www.agroecology-europe.org/wp-content/uploads/2020/11/AEEU-Mapping-Report-agroecological-practices-November-version.pdf

AEEU (2021) *The Agroecology Europe Barcelona Letter*, Corbais: Agroecology Europe (AEEU), www.agroecology-europe.org/the-agroecology-europe-barcelona-letter

AEEU (2022) *Improving Eco-Schemes in the Light of Agroecology: Key Recommendations for the 2023–2027 Common Agricultural Policy*, Corbais: Agroecology Europe (AEEU), www.agroecology-europe.org/wp-content/uploads/2022/02/Improving-eco-schemes-in-the-light-of-agroecology-Policy-Brief-Feb-2022-AE4EU.pdf

AER and FoEE (2005) *Safeguarding Sustainable European Agriculture: Coexistence, GMO Free Zones and the Promotion of Quality Food Produce in Europe*, 17 May conference, Assembly of European Regions (AER) and Friends of the Earth Europe (FoEE), www.a-e-r.org

AFL-CIO Energy Committee (2019) *Letter to US Representatives Alexandria Ocasio-Cortez and Ed Markey*, 8 March, Washington, DC: AFL-CIO Energy Committee, www.ibew.org/Portals/22/IBEW%20Letters/2019/Markey. Ocasio-Cortez%20Letter.Climate.pdf?ver=2019-03-11-111951-970

Air Products (2011) Air Products submits planning application for Tees Valley Renewable Energy Facility; facility will turn waste into electricity for 50,000 homes in the North East, *Air Products*, 15 February, www.airp roducts.co.uk/company/news-center/2011/02/0215-air-products-subm its-planning-application-for-tees-valley-renewable-energy-facility.aspx

Air Products (2016) Air Products will exit energy-from-waste business, *Air Products*, 4 April, www.airproducts.co.uk/company/news-center/2016/ 04/0404-air-products-will-exit-energy-from-waste-business

Ajl, M. (2021) *A People's Green New Deal*, London: Pluto, https://library. oapen.org/handle/20.500.12657/48775

Alexander, S. (2015) *Sufficiency Economy: Enough, for Everyone, Forever*, Melbourne: Simplicity Institute.

Alexander, C. and Reno, J. (2014) From biopower to energopolitics in England's modern waste technology, *Anthropological Quarterly* 87(2): 335–358.

Allaire, G. and Wolf, S. (2004) Cognitive representations and institutional hybridity in agrofood innovation, *Science, Technology and Human Values* 29(4): 431–458.

Allin, S. (2015) The future of waste to energy technology, Business Development Director Sales at Babcock & Wilcox Volund, in Essays for EfW conference.

APP (2013) *Converting Waste into Valuable Resources with the Gasplasma® Process*, presentation to 2013 Gasification Technologies Conference by Chief Executive, Advanced Plasma Products (APP), www.gasificat ion.org/uploads/eventLibrary/2013-11-3-Stein-Advanced-Plasma-Power.pdf

ARC2020 (2010) A Communication from Civil Society to the European Union Institutions on the future Agricultural and Rural Policy, Brussels, www.arc2020.eu/communication/

ARC2020 (2013) *CAP Deal: Weak – But Options Worth Fighting For*, Brussels: Agricultural and Rural Convention (ARC), www.arc2020.eu

ARC2020 (2022) *Rural Europe Takes Action: No More Business as Usual.* Brussels, www.arc2020.eu/tag/rural-europe-takes-action/

ARC2020, FoEE and IFOAM EU (2015) *Transitioning Towards Agroecology: Using the CAP to Build New Food Systems*, Agricultural and Rural Convention 2020 (ARC2020), Friends of the Earth Europe (FoEE) and International Federation of Organic Agriculture Movements (IFOAM EU), www.arc2020.eu/wp-content/uploads/2015/02/arc2020-brochure-5-with-hyperlinks.pdf

ARC2020, IFOAM EU and TP Organics (2012) *Agro-Ecological Innovation project: Progress and Recommendations*, Agricultural and Rural Convention 2020 (ARC2020), International Federation of Organic Agriculture Movements (IFOAM EU) and Technology Platform Organics (TP Organics), http://agro-ecoinnovation.eu/toolbox/july-2012-workshop-materials/

Arena, U. (2011) Gasification: An alternative solution for waste treatment with energy recovery, *Waste Management* 31: 405–406.

Aronoff, K. (2021) The big difference between a Green New Deal and Biden's climate agenda, *New Republic*, 20 April, https://newrepublic.com/article/162106/big-difference-green-new-deal-bidens-climate-agenda

Aronoff, K. (2022) To fight climate change, bring back state planning, *Intelligencer*, 26 January, https://nymag.com/intelligencer/2022/01/to-fight-climate-change-bring-back-state-planning.html

Aronoff, K., Battistoni, A., Cohen, D.A. and Riofrancos, T. (2019) *A Planet to Win: Why We Need a Green New Deal*, New York: Verso.

ASEED (2008) Agrofuel, no cure for oil addiction and climate change, https://aseed.net/brochure-uagrofuel-no-cure-for-oil-addiction-and-climate-changeu/

Bailey, I. (2007) Market environmentalism, new environmental policy instruments, and climate change policy in the United Kingdom and Germany, *Annals of the Association of American Geographers* 97(3): 530–550.

Bailey, R. (2008) *Another Inconvenient Truth: How Biofuel Policies are Deepening Poverty and Accelerating Climate Change*, Oxford: Oxfam International.

Baker, S. (2007) Sustainable development as symbolic commitment: Declaratory politics and the seductive appeal of ecological modernisation in the European Union, *Environmental Politics* 16(2): 297–317.

Barroso, J.M. (2007) President of the European Commission, keynote speech on Biofuels, International Biofuels Conference, Brussels, 5 July, Speech/07/470.

Barry, J. and Paterson, M. (2004) Globalisation, ecological modernisation and New Labour, *Political Studies* 52: 767–784.

BBC (2012) Deal signed on £500m incinerator for Gloucester, *BBC*, 12 September, www.bbc.co.uk/news/uk-england-gloucestershire-19571080

BEIS (2018) *A Future Framework for Heat in Buildings: Government Response*, London: Department of Business, Energy and Industrial Strategy (BEIS).

BEIS (2021) *Heat and Buildings Strategy*, London: Department for Business, Energy and Industrial Strategy (BEIS).

BEIS Committee (2022) *Decarbonising Heat in Homes*, London: House of Commons Business, Energy and Industrial Strategy (BEIS) Committee, HC 1038.

Bernstein, S. (2001) *The Compromise of Liberal Environmentalism*, New York: Columbia University Press.

Beuret, N. (2019) A Green New Deal between whom and for what?, *Viewpoint*, 24 October, viewpointmag.com/2019/10/24/green-new-deal-for-what

Beuret, N. (2021) The green transition is already underway – and it's not looking pretty, *Novara Media*, 22 October, https://novaramedia.com/2021/10/22/the-green-transition-is-already-underway-and-its-not-looking-pretty/

Bhattacharya, T. (2019) Three ways a Green New Deal can promote life over capital, *Jacobin*, 6 October, https://jacobinmag.com/2019/06/green-new-deal-social-care-work

Biden Climate Plan (2020) *Biden Plan for a Clean Energy Revolution and Environmental Justice*, Washington, DC: Democratic National Committee, https://joebiden.com/climate-plan/

Bijker, W.E. (1997) *Of Bicycles, Bakelites, and Bulbs: Toward a Theory of Sociotechnical Change*, Cambridge, MA: MIT Press.

Bioenergy Out Declaration (2015) Declaration: Large-scale bioenergy must be excluded from the renewable energy definition, www.econexus.info/files/BioenergyOut-Declaration-and-signatories.pdf

Biofrac (2006) *Biofuels in the European Union: A Vision for 2030 and Beyond*, final draft report of the Biofuels Research Advisory Council, https://etipbioenergy.eu/images/biofuels_vision_2030.pdf

Birch, K., Levidow, L. and Papaioannou, T. (2010) Sustainable capital? The neoliberalization of nature and knowledge in the European 'Knowledge-Based Bio-economy', *Sustainability* 2(9): 2898–2918.

BirdLife International et al (2010) *Driving to Destruction: The Impacts of Europe's Biofuel Plans on Carbon Emissions and Land*, commissioned by ActionAid, BirdLife International, ClientEarth, European Environmental Bureau, FERN, Friends of the Earth Europe, Greenpeace, Transport & Environment, Wetlands International.

Bollier, D. and Helfrich, S. (2015) *Patterns of Commoning*, The Commons Strategy Group, https://patternsofcommoning.org

Bond, P. (2012) *Politics of Climate Justice: Paralysis Above, Movement Below*, Scottsville: University of KwaZulu-Natal Press.

Bond, P. (2018) Climate Justice during the decline of global governance, in Lele, S., Brondizio, E.S., Byrne, J., Mace, J.M. and Martinez-Alier, J. (eds) *Rethinking Environmentalism: Linking Justice, Sustainability and Diversity*, Cambridge, MA: MIT Press, pp 153–182.

Borras, J., Franco, J., Gómez, S., Kay, C. and Spoor, M. (2012) Land grabbing in Latin America and the Caribbean, *Journal of Peasant Studies* 39(3–4): 845–872.

Borras, S., McMichael, P. and Scoones, I. (2010) The politics of biofuels, land and agrarian change: editors' introduction, *Journal of Peasant Studies* 37(4): 575–592.

Bowers, C. (1993) Europe's motorways, *The Ecologist* 23(4): 125–130.

Bowyer, C. (2010) *Anticipated Indirect Land Use Change Associated with Expanded Use of Biofuels and Bioliquids in the EU: An Analysis of the National Renewable Energy Action Plans*, IEEP, www.ieep.eu

Boyle, M., McWilliams, C. and Rice, G. (2008) The spatialities of actually existing neoliberalism in Glasgow, 1977 to present, *Geografiska Annaler: Series B, Human Geography*, 90(4): 313–325.

Bradbury, H. (2010) What is good action research?: Why the resurgent interest?, *Action Research* 8: 93–109.

Bruno, K., Karliner, J. and Brotsky, C. (1999) Greenhouse gangsters vs. climate justice, *Corporate Watch*, www.corpwatch.org/article.php?id=1048

Buller, A. (2020) Where next for the Green New Deal?, *Renewal* 28(1), https://lwbooks.co.uk/product/where-next-for-the-green-new-deal-rene wal-1-spring-2020

Busch, L. (2010) Can fairy tales come true? The surprising story of neoliberalism and world agriculture, *Sociologia Ruralis* 50(4): 331–351.

Business Europe (2009) *Priorities in the Field of Transport 2009–2014*, www. businesseurope.eu/publications/businesseuropes-priorities-field-transp ort-2009-2014

Buttel, F. (2005) The environmental and post-environmental politics of GM crops and foods, *Environmental Politics* 14(3): 309–323.

Buttel, F.H. (2000) Ecological modernization as social theory, *Geoforum* 31: 57–65.

CACCTU (2021) *Climate Jobs: Building a Workforce for the Climate Emergency*, London: Campaign Against Climate Change Trade Union Group (CACCTU), www.cacctu.org.uk/climatejobs

Caffentzis, G. (2016) Commons, in Fritsch, K., O'Connor, C. and Thompson, A.K., (eds), *Keywords for Radicals: The Contested Vocabulary of Late-Capitalist Struggle*, Chicago and Edinburgh: AK Press, pp 95–102.

CAN (2017) *EU's Emissions Trading Scheme*, Climate Action Network (CAN) European NGO Coalition, www.caneurope.org/climate/emissi ons-trading-scheme

Carbon Market Watch (2014) Not smart: Climate smart agriculture in carbon markets, *Carbon Market Watch*, 25 November, https://carbonmark etwatch.org/2014/11/25/promoting-climate-smart-agriculture-with-car bon-markets-would-not-be-a-smart-move/

Carrington, D. (2013) Angela Merkel 'blocks' EU plan on limiting emissions from new cars, *The Guardian*, 28 June, www.theguardian.com/environm ent/2013/jun/28/angela-merkel-eu-car-emissions

Carroll, L. (1871) *Through the Looking-Glass*, London: Macmillan.

Carson, R. (1962) *Silent Spring*, New York: Fawcett Crest.

Carton, W. (2019) 'Fixing' climate change by mortgaging the future: Negative emissions, spatiotemporal fixes, and the political economy of delay, *Antipode* 51(3): 750–769.

CEC (1993a) Growth, competitiveness, employment: The challenges and ways forward into the 21st century, *Bulletin of the European Communities*, supplement 6/93 [especially pp 100–103], Brussels: Commission of the European Communities (CEC).

CEC (1993b) *Towards Sustainable Development*, Fifth Environmental Action Programme; also in *Official Journal of the European Communities*, C 138, 17 May, pp 5–98.

CEC (2003) Commission recommendation of 23 July on guidelines for the development of national strategies and best practices to ensure co-existence of GM crops with conventional and organic farming.

CEC (2006a) *An EU Strategy for Biofuels*, Communication from the Commission, SEC(2006) 142.

CEC (2006b) Commission Staff Working Document – Annex to the Communication from the Commission – An EU Strategy for Biofuels – Impact Assessment COM(2006) 34 final, SEC(2006) 0142.

CEC (2007a) *An Energy Policy for Europe*, Communication from the Commission to the European Council and the European Parliament.

CEC (2007b) *Biofuels Progress Report: Report on the Progress Made in the Use of Biofuels and Other Renewable Fuels in the Member States of the European Union*, SEC(2006) 1721.

CEC (2008a) European Commission proposes new strategy to address EU critical needs for raw materials, 4 November, IP(2008) 1628.

CEC (2008b) Commission Staff Working Paper accompanying the Communication, *The EU: A Global Partner for Development: Speeding Up Progress towards the Millennium Development Goals*, DGs Climate Change, Energy, Biofuels, Migration and Research, COM(2008) 177 final.

CEC (2010a) *Europe 2020: A Strategy for Smart, Sustainable and Inclusive Growth*, Brussels: Commission of the European Communities (CEC).

CEC (2010b) *Europe 2020 Flagship Initiative: Innovation Union, SEC (2010) 1161*, Brussels: Commission of the European Communities (CEC).

CEC (2010c) Report from the Commission on indirect land-use change related to biofuels and bioliquids, COM(2010) 811 final.

CEC (2012) Proposal for a Directive amending Directive 2009/28 of the European Parliament and of the Council of 23 April 2009 on the promotion of the use of energy from renewable sources, 11 September (document for internal consultation among Commission Services), www.foeeurope.org/ biofuels-reform-messy-compromise-120912

CEC (2019) *The European Green Deal*, Brussels, 11.12.2019 COM(2019) 640 final, Brussels: Commission of the European Communities (CEC).

CEC-Innovation (2019) New Entrants' Reserve (NER) programme. Co-funded by the NER programme of the European Union, Brussels: Commission of the European Communities, https://ec.europa. eu/clima/policies/innovation-fund/ner300_en

CEO (2007) *The EU's Agrofuel Folly*, 4 June, Brussels: Corporate Europe Observatory (CEO), https://corporateeurope.org/en/food-and-agricult ure/2007/06/eus-agrofuel-folly

CEO (2008) *Global Europe: An Open Door Policy for Big Business Lobbyists at DG Trade*, October, Brussels: Corporate Europe Observatory (CEO).

CEO (2009) *Car Industry Flexes its Muscles, Commission Bows Down*, Briefing Paper, 20 January, Brussels: Corporate Europe Observatory (CEO).

CEO (2015) *EU Emissions Trading: 5 Reasons to Scrap the ETS*, 26 October, Brussels: Corporate Europe Observatory (CEO), https://corporat eeurope.org/en/environment/2015/10/eu-emissions-trading-5-reas ons-scrap-ets

CEO (2020) *The Hydrogen Hype: Gas Industry Fairy Tale or Climate Horror Story?* Brussels: Corporate Europe Observatory (CEO).

CEO, GRR, Econexus and Biofuelwatch (2008) Joint press release: Sustainability criteria and certification of biomass – greenwashing destruction in pursuit of profit.

CEO, TNI and GRR (2007) *Paving the Way for Agrofuels EU Policy: Sustainability Criteria and Climate Calculations*, Amsterdam: Transnational Institute.

CETRI (2010) *Agrocarburants: Impacts au Sud*, Louvain-la-Neuve: Le Centre Tricontinental (CETRI), www.cetri.be/spip.php?rubrique132

Chataway, J., Tait, J. and Wield, D. (2004) Understanding company R&D strategies in agro-biotechnology: trajectories and blind spots, *Research Policy* 33(6–7): 1041–1057.

Chatterton, P., Featherstone, D. and Routledge, P. (2013) Articulating climate justice in Copenhagen: Antagonism, the commons, and solidarity, *Antipode* 45(3): 602–620.

Christophers, B. (2022) Fossilised capital: Price and profit in the energy transition, *New Political Economy* 27(1): 146–159, www.tandfonline.com/ journals/cnpe20

CIDSE (2015) Climate-smart revolution … or a new era of green washing?, *CIDSE*, www.cidse.org/wp-content/uploads/2015/05/753_751_750_ CIDSE_GACSA_briefing_FINAL.pdf

Cintron, I. (2015) Lobbyists at COP21: Foxes guarding the henhouse?, *DW Global Media Forum*, 12 December, www.dw.com/en/lobbyists-at-cop21- foxes-guarding-the-henhouse/a-18902450

Ciplet, D. and Roberts, J.T. (2017) Climate change and the transition to neoliberal environmental governance, *Global Environmental Change* 46: 148– 156, www.mdpi.com/2076-3298/4/4/73/pdf

CIWM (2013) *Technical Briefing Note: Thermal Treatment of Waste*, London: Chartered Institution of Wastes Management (CIWM).

CIWM (2014) *The Circular Economy: What Does It Mean for the Waste and Resource Management Sector?* London: Chartered Institute of Waste Management (CIWM).

Clare, D. (2019) Carbon markets will not help stop climate change, *Climate Home News*, 13 December, www.climatechangenews.com/2019/12/13/carbon-markets-will-not-help-stop-climate-change

CMW (2018) *The Clean Development Mechanism: Local Impacts of a Global System*, Carbon Market Watch (CMW).

Coghlan, D. and Brydon, M. (eds) (2014) *The SAGE Encyclopedia of Action Research*, Thousand Oaks, CA, USA.

Common Weal (2019) *Our Common Home: A Green New Deal for Scotland*, Common Weal, https://commonweal.scot/index.php/building-green-new-deal-scotland

Common Weal (2020a) *The Common Home Plan: Homes and Buildings*, Common Weal, https://commonweal.scot/our-common-home/homes-buildings

Common Weal (2020b) *The Common Home Plan: Heating*, Common Weal, https://commonweal.scot/our-common-home/heating

Common Weal et al (2019) *The Future of Low Carbon Heat for Off-Gas Buildings: A Call for Evidence*, Glasgow: Common Weal, Glasgow Caledonian University, and the Energy Poverty Research Initiative, https://commonweal.scot/index.php/policy-library/future-low-carbon-heat-gas-buildings

Community R4C (2015) Animated cartoon describing an alternative to the Javelin Park incinerator, *Community R4C*, www.youtube.com/watch?v=kf3IjfYKZg4

Connett, P. (2013) *The Zero Waste Solution: Untrashing the Planet One Community at a Time*, Chelsea Green Publishing, www.chelseagreen.com/the-zero-waste-solution

COPA-COGECA (2013a) *The Common Agricultural Policy after 2013*, Brussels: Committee of Professional Agricultural Organisations (COPA), www.copa-cogeca.be

COPA-COGECA (2013b) *Copa-Cogeca's Position on the EU's Biofuels Policy*. Brussels: Organizational author.

Corbey, D. (2007) Draft report (PE392.119v01-00). Proposal for a directive of the European Parliament and of the Council amending Directive 98/70/EC as regards the specification of petrol, diesel and gas-oil and introducing a mechanism to monitor and reduce greenhouse gas emissions from the use of road transport fuels, *European Parliament*, 11 October.

Corporate Watch (2008) *Technofixes: A Critical Guide to Climate Change Technologies*, Corporate Watch, https://corporatewatch.org/product/technofixes-a-critical-guide-to-climate-change-technologies/

Corporate Watch (2019) The zombie technofix, *The Ecologist*, 25 January, https://theecologist.org/2019/jan/25/zombie-technofix

Corvellec, H. and Hultman, J. (2011) *From 'a Farewell to Landfill' to 'a Farewell to Wastefulness': Societal Narratives, Socio-materiality and Organizations*, Working Paper No 1, Lund University, Research in Service Studies.

Corvellec, H. and Hultman, J. (2012) From less landfilling' to 'wasting less': Societal narratives, socio-materiality and organizations, *Journal of Organizational Change Management* 25(2): 297–314.

Cox, E., Johnstone, P. and Stirling, A. (2016) *Understanding the Intensity of UK Policy Commitments to Nuclear Power*, SPRU Working Paper Series, www.sussex.ac.uk/spru/newsandevents/2017/findings/nuclear

Cozzens, S.E. and Kaplinsky, R. (2009) Innovation, poverty and inequality: cause, coincidence, or co-evolution?, in B.-A. Lundvall, J.K. Joseph, C. Chaminade and J. Vang (eds) *Handbook of Innovation Systems and Developing Countries: Building Domestic Capabilities in a Global Context*, Cheltenham: Edward Elgar, https://doi.org/10.4337/9781849803 427.00009

CPE (2001) *To Change the CAP*, Brussels: Coordination Paysanne Européenne (CPE) which later became Via Campesina European Coordination.

Creech, L. (2017) Residual waste treatment expansion could set ceiling on UK recycling, *Resource*, 7 August, https://resource.co/article/residual-waste-treatment-expansion-could-set-ceiling-uk-recycling-12014

CropLife (2014) *Save Our Soil*, Brussels: CropLife International, https://croplife.org/news/save-our-soil/

CSA Concerns (2014) Open letter from civil society on the Global Alliance for Climate-Smart Agriculture (CSA), www.climatesmartagconcerns.info/open-letter.html

CXC (2022) *Clean Heat and Energy Efficiency Workforce Assessment*, Edinburgh: ClimateXChange (CXC), Clean Heat and Energy Efficiency Workforce Assessment.

Dale, G. (ed) (2016) *Green Growth: Political Ideology, Political Economy and Policy Alternatives*, London: Zed.

Danish Government (2013) *Denmark without Waste: Recycle More, Incinerate Less*, Copenhagen: Dakofa, https://dakofa.com/element/denmark-without-waste-recycle-more-incinerate-less/.

de Angelis, M. (2003) Commons and communities: Or building a new world from the bottom up, *The Commoner* 6, www.commoner.org.uk/deangelis06.pdf

de Angelis, M. (2017) *Omnia Sunt Communia: On the Commons and the Transformation to Postcapitalism*, London: Bloomsbury.

DECC (2012) *Government Response to the Consultation on Proposals for the Levels of Banded Support under the Renewables Obligation for the Period 2013–17 and the Renewable Obligation Order 2012*, London: Dept for Energy and Climate Change (DECC).

DECC (2014) Consultation on changes to grandfathering policy with respect to future biomass co-firing and conversion projects in the Renewables Obligation.

DEFRA (2007a) *UK Biomass Strategy*, London: Department for Environment, Food and Rural Affairs (DEFRA), www.globalbioenergy.org/uploads/media/0705_Defra_-_UK_Biomass_Strategy_01.pdf

DEFRA (2007b) *Waste Strategy for England 2007*, London: Department for Environment, Food and Rural Affairs (DEFRA), www.gov.uk/governm ent/uploads/system/uploads/attachment_data/file/228536/7086.pdf

DEFRA (2011a) *Applying the Waste Hierarchy: Evidence Summary*, London: Department for Environment, Food and Rural Affairs (DEFRA).

DEFRA (2011b) *The Economics of Waste and Waste Policy*, London: Department for Environment, Food and Rural Affairs (DEFRA), https://assets.publish ing.service.gov.uk/government/uploads/system/uploads/attachment_data/ file/69500/pb13548-economic-principles-wr110613.pdf

DEFRA (2012) *Local Authority Collected Waste Management Statistics for England: Final Annual Results 2011/12*, London: Department for Environment, Food and Rural Affairs (DEFRA), https://webarchive.natio nalarchives.gov.uk/20130222092708/http://www.defra.gov.uk/statistics/ files/mwb201112_statsrelease.pdf

DEFRA (2013a) *Incineration of Municipal Solid Waste*, London: Department for Environment, Food and Rural Affairs (DEFRA).

DEFRA (2013b) *Advanced Thermal Treatment of Municipal Solid Waste*, London: Department for Environment, Food and Rural Affairs (DEFRA).

DEFRA (2014a) *Energy from Waste: A Guide to the Debate*, February (rev. edn), London: Department for Environment, Food and Rural Affairs (DEFRA), www.gov.uk/government/uploads/system/uploads/attachme nt_data/file/284612/pb14130-energy-waste-201402.pdf

DEFRA (2014b) *Statistics on Waste Managed by Local Authorities in England in 2012/13*, London: Department for Environment, Food and Rural Affairs (DEFRA), www.gov.uk/government/uploads/system/uploads/attachme nt_data/file/255610/Statistics_Notice1.pdf

DEFRA (2018) *Our Waste, Our Resources: A Strategy for England*, London: Department for Environment, Food and Rural Affairs (DEFRA), https://assets.publishing.service.gov.uk/government/uploads/system/uplo ads/attachment_data/file/765914/resources-waste-strategy-dec-2018.pdf

Deininger, K. and Byerlee, D. (2010) *Rising Global Interest in Farmland: Can It Yield Sustainable and Equitable Benefits?* Washington, DC: World Bank.

della Porta, D. and Parks, L. (2017) Framing processes in the climate movement: From climate change to climate justice, in M. Dietz and H. Garrelts (eds) *Routledge Handbook of Climate Change Movements*, London: Routledge, www.academia.edu/3531742/Framing_Processes_ in_the_Climate_Movement_from_Climate_Change_to_Climate_Justice.

della Porta, D. and Portos, M. (2021) Rich kids of Europe? Social basis and strategic choices in the climate activism of Fridays for Future, *Italian Political Science Review/Rivista Italiana di Scienza Politica*, doi:10.1017/ipo.2021.54

DG Agri (2007) *The Impact of a Minimum 10% Obligation for Biofuel Use in the EU-27 in 2020 on Agricultural Markets*, Brussels: Directorate-General for Agriculture and Rural Development, European Commission.

DG Research (2005) *New Perspectives on the Knowledge-Based Bio-Economy: Conference Report*, Brussels: DG Research.

Di Chiro, G. (1998) Nature as community: The convergence of social and environmental justice, in M. Goldman (ed) *Privatizing Nature: Political Struggles for the Global Commons*, London: Pluto, pp 298–320.

Dodds, L. and Hopwood, B. (2006) Ban waste, environmental justice and citizen participation in policy setting, *Local Environment* 11(3): 269–286.

Dontenville, A. (2009) *The Governance Regime of Biofuels Risk-Risk Trade-Offs in the European Union: An Explanatory Study of Factors Shaping Regulatory Content*, MA dissertation, King's College London.

Drechsler, W. (2011) *Techno-Economic Paradigms: Essays in Honour of Carlota Perez*, London: Anthem Press.

Dyke, J., Watson, J. and Knorr, W. (2021) Climate scientists: Concept of net zero is a dangerous trap, *The Conversation*, 22 April, https://theconversation.com/climate-scientists-concept-of-net-zero-is-a-dangerous-trap-157368

EAC (2014) *Growing a Circular Economy: Ending the Throwaway Society*, London: Environmental Audit Committee (EAC), House of Commons.

EBTP (2008) *European Biofuels Technology Platform: Strategic Research Agenda & Strategy Deployment Document*, CPL Press.

EBTP (2015) EBTP Position on the Rapporteur's amendments to the review of the RED and FQD, Brussels: European Biofuels Technology Platform.

EC (1998) Directive 98/44/EC of the European Parliament and of the Council on Protection of Biotechnological Inventions, *Official Journal of the European Union*, 30 July, L 213: 13.

EC (2000) Directive 2000/76/EC on the incineration of waste ('WID'), *Official Journal of the European Communities*, L 332/91.

EC (2001) European Parliament and Council Directive 2001/18/EC of 12 March on the deliberate release into the environment of genetically modified organisms and repealing Council Directive 90/220/EEC, *O.J.* L 106: 1–38.

EC (2003a) Regulation 1829/2003 of 22 September 2003 on genetically modified food and feed, *O.J.* L 268, 18 October: 1–23.

EC (2003b) Regulation 1830/2003 of 22 September 2003 concerning the traceability and labelling of GMOs and traceability of food and feed produced from GMOs and amending Directive 2001/18, *O.J.* L 268, 18 October: 24–28.

EC (2008) Directive 2008/98/EC of the European Parliament and of the Council of 19 November 2008 on waste and repealing certain Directives, *O.J.* L 312/30–30, http://ec.europa.eu/environment/waste/framework/, http://eur-lex.europa.eu/legal-content/EN/TXT/?uri=CELEX:32008L0098

EC (2009) Directive 2009/28/EC of the European Parliament and of the Council of 23 April 2009 on the promotion of the use of energy from renewable sources and amending and subsequently repealing Directives 2001/77/EC and 2003/30/EC Renewable Energy Directive, *O.J.* L 140: 16–62, 5 June.

EC (2010) Directive 2010/75/EC on Industrial Emissions (integrated pollution prevention and control), *O.J.* L 334/17–119, http://eur-lex.eur opa.eu/LexUriServ/LexUriServ.do?uri=OJ:L:2010:334:0017:0119:en:PDF

EC (2014) *Towards a Circular Economy: A Zero Waste Programme for Europe* {SWD(2014) 206 final} {SWD(2014) 211 final}

EC (2015) *Closing the Loop: An EU Action Plan for the Circular Economy*, https:// eur-lex.europa.eu/legalcontent/EN/TXT/?uri=CELEX:52015DC0614

EC (2017) The role of waste-to-energy in the circular economy, European Commission, 26 January, https://eur-lex.europa.eu/legal-content/en/ TXT/?uri=CELEX%3A52017DC0034

EC (2018) Directive 2018/2001/EU of the European Parliament and of the Council of 11 December 2018, https://ec.europa.eu/jrc/en/jec/renewa ble-energy-recast-2030-red-ii

Econexus (2007) *Call for an immediate moratorium on EU incentives for agrofuels, EU imports of agrofuels and agroenergy monocultures*, www.econexus.info/ call-immediate-moratorium-eu-incentives-agrofuels-eu-imports-agrofu els-and-eu-agroenergy-monocultur-0

Econexus et al (2007) *Agrofuels: Towards a Reality Check in Nine Key Areas*, Biofuelwatch, Carbon Trade Watch, Transnational Institute, Corporate Europe Observatory, EcoNexus, Ecoropa, Grupo de Reflexión Rural, Munlochy Vigil, NOAH (Friends of the Earth Denmark), Rettet den Regenwald, Watch Indonesia.

Econexus et al (2009) *Agriculture and Climate Change: Real Problems, False Solutions*, Grupo de Reflexión Rural, Biofuelwatch, EcoNexus, NOAH & FoE Denmark, www.econexus.info/files/agriculture-climate-change-june-2009_summary.pdf.

Ecosocialist Encounter (2022) *5th International Ecosocialist Encounter: Final Report*, https://scote3.files.wordpress.com/2022/05/5th-ecosoc-final-report-v6.pdf

EcoWatch (2020) Greta Thunberg warns humanity 'Still speeding in wrong direction' on climate, *EcoWatch*, 11 December, www.ecowatch.com/greta-thunberg-paris-agreement-2649450989.html

ECVC (2015) *We Want System Change Not Climate Change*, Brussels: European Coordination Via Campesina (ECVC), www.eurovia.org/we-want-system-change-not-climate-change/

ECVC (2020) *Farm to Fork Strategy: Key Messages From ECVC*, Brussels: European Coordination Via Campesina (ECVC), www.eurovia. org/wp-content/uploads/2020/02/F2F-ECVC-EN.pdf

EDF (2007) *Harvesting the Low-Carbon Cornucopia: How the European Union Emissions Trading System (EU-ETS) is Spurring Innovation and Scoring Results, by Annie Petsonk and Jos Cozijnsen*, Washington, DC: Environmental Defense Fund (EDF), https://cprubibliography.wordpress.com/2012/07/18/harvest ing-the-low-carbon-cornucopia-how-the-european-union-emissions-trad ing-system-eu-ets-is-spurring-innovation-and-scoring-results/

Edgerton, D. (2006) *The Shock of the Old: Technology and Global History since 1900*. Oxford: Oxford University Press.

EEA (2009) *Guidelines on Waste Prevention Programmes*, Copenhagen: European Environment Agency (EEA), http://eea.europa.eu/themes/waste

EEA (2020) *Average Greenhouse Gas Intensity of Road Transport Fuels in the EU, 2010–2018*, Copenhagen: European Environment Agency (EEA), www.eea.europa.eu/data-and-maps/figures/average-greenhouse-gas-intensity-of

EEC (1990) Council Directive 90/220/EEC on the deliberate release to the environment of genetically modified organisms, *Official Journal of the European Communities*, L 117: 15–27.

EESC (2019) *European Agriculture Should Develop Towards Agroecology*, Brussels: European Economic and Social Committee (EESC), www.eesc.europa.eu/en/news-media/news/european-agriculture-should-develop-towards-agroecology-0

Eide, A. (2008) *The Right to Food and the Impact of Liquid Biofuels (Agrofuels)*, Rome: Right to Food Studies, FAO.

EMF (2013) *Towards the Circular Economy: Economic and Business Rationale for an Accelerated Transition*, Ellen MacArthur Foundation, www.ellenmacarthurfoundation.org/assets/downloads/publications/Ellen-MacArthur-Foundation-Towards-the-Circular-Economy-vol.1.pdf

Eminton, S. (2021) Activists set for EfW day of action, *Let's Recycle*, 21 September, www.letsrecycle.com/news/activists-set-for-efw-day-of-action/

Empson, M. (ed) (2019) *System Change Not Climate Change: A Revolutionary Response to Environmental Crisis*, London: Bookmarks.

Energos (2016) Statement of administrator's proposals, 13 September, https://beta.companieshouse.gov.uk/company/03109022/filing-history

Ernsting, A. and Smolker, R. (2018) *Dead End Road: The False Promises of Cellulosic Biofuels*, Biofuelwatch, www.biofuelwatch.org.uk/wp-content/uploads/Cellulosic-biofuels-report-2.pdf

ESA (2018) *Energy for the Circular Economy: An Overview of Energy from Waste in the UK*, London: Environmental Services Association (ESA), www.esauk.org/application/files/7715/3589/6450/20180606_Energy_for_the_circular_economy_an_overview_of_EfW_in_the_UK.pdf

ETC Group (2015) Time to wave the white flag for a failed techno-fix?, *ETC Group*, www.etcgroup.org/content/time-wave-white-flag-failed-techno-fix

ETI (2012) ETI seeks partners for £13 million Energy from Waste demonstrator plant, *ETI*, 9 April, www.energytechnologies.co.uk/Home/Technology-Programmes/BioEnergy.aspx

EUA (2021) *Decarbonising Heat in Buildings: Putting Consumers First*, London: Energy Union Alliance (EUA), https://eua.org.uk/uploads/608167B5BC925.pdf

EU ETS (2005) *EU Emissions Trading System (EU ETS)*, https://ec.europa. eu/clima/policies/ets_en

EU ETS (2015) *EU ETS Handbook*, Brussels: European Commission, https://aeaep.com.ua/en/wp-content/uploads/2015/07/ets_handbook _en.pdf

Eunomia (2006) *A Changing Climate for Energy from Waste?*, Bristol: Eunomia, www.no-burn.org/wp-content/uploads/changing_climate.pdf

Eunomia (2017) *Residual Waste Infrastructure Review (12th Issue)*, Bristol: Eunomia, www.eunomia.co.uk/reports-tools/residual-waste-inf rastructure-review-12th-issue/

Euractiv (2008) Biofuels for transport, *Euractiv*, 11 April, www.euractiv. com/en/transport/biofuels-transport-linksdossier-188374

EuropAfrica (2012) *Bio-fuelling Injustice? EuropAfrica 2011 Monitoring Report on EU Policy Coherence for Food Security*, Rome: Terra Nuova, www.terranu ova.org/publications/biofueling-injustice-2012.

EuropeAid (2009) *Position on Biofuels for the ACP-EU Energy Facility*, Brussels: Directorate-General for Development and Cooperation, http:// web.archive.org/web/20131104020818/http://ec.europa.eu/europeaid/ where/acp/regional-cooperation/energy/documents/biofuels_position_ paper_en.pdf.

Evans, G. (2014) *ETI Waste Gasification Project*. Dr Geraint Evans, Programme Manager – Bioenergy. Energy Technologies Institute, https://s3-eu-west- 1.amazonaws.com/assets.eti.co.uk/legacyUploads/2014/10/14-10-29- GDE-ETI-gasification-presentation-IEA-Harwell-short-comms-UPDA TED.pdf

Fairlie, S. (1993) The infrastructure lobby, *The Ecologist* 23(4): 123–124.

Fals Borda, O. (2001) Participatory action research in social theory: Origins and challenges, in P. Reason and H. Bradbury (eds) *Handbook of Action Research: Participatory Inquiry and Practice*, Thousand Oaks: SAGE, pp 27–37.

FAO (2013) *Climate-Smart Agriculture Sourcebook*, Rome: Food and Agriculture Organization (FAO), www.fao.org/3/i3325e/i3325e.pdf

FAO (2016) *Recommendations from the Participants*, Regional Symposium on Agroecology for Sustainable Agriculture and Food Systems in Europe and Central Asia, 23–25 November 2016, Budapest.

Farand, C. (2018) Trade union lobbying risks slowing down transition to zero-carbon future, *Desmog*, 27 September, www.desmog.co.uk/2018/ 09/27/trade-union-lobbying-risks-slowing-down-transition-zero-car bon-future

Fargione, J., Hill, J., Tilman, D., Polasky, S. and Hawthorne, P. (2008) Land clearing and the biofuel carbon debt, *Science*, 319(5867): 1235–1238.

Featherstone, D., Ince, A., Mackinnon, D. and Strauss, K. (2012) Progressive localism and the construction of political alternatives, *Transactions of the Institute of British Geographers* 37: 177–182.

Felt, U., Wynne, B., Callon, M., Gonçalves, M.E., Jasanoff, S., Jepsen, M., et al (2007) *Taking European Knowledge Society Seriously*. Report of the Expert Group on Science and Governance to the Science, Economy and Society Directorate, Directorate-General for Research. Luxembourg: Office for Official Publications of the European Communities.

Feola, G. and Butt, A. (2017) The diffusion of grassroots innovations for sustainability in Italy and Great Britain: An exploratory spatial data analysis, *Geographical Journal* 183(1): 16–33.

Feola, G. and Jaworska, S. (2019) One transition, many transitions? A corpus-based study of societal sustainability transition discourses in four civil society's proposals, *Sustainability Science* 14: 1643–1656.

Fernandez, M., Goodall, K., Olson, M. and Méndez, V.E. (2013) Agroecology and alternative agri-food movements in the United States: Toward a sustainable agri-food system, *Agroecology and Sustainable Food Systems* 37(1): 115–126.

FFA (2005) *Berlin Manifesto for GMO-free Regions and Biodiversity in Europe*, Foundation for Future Farming (FFA), Assembly of European Regions, January, www.are-regions-europe.org

FfF (2020) *Change the CAP! Open Letter to the EU*, 22 April, Fridays for Future (FfF), https://fridaysforfuture.org/change-the-cap/

FIAN International (2008) *Agrofuels in Brazil: Report of the Fact-finding Mission on the Impacts of Public Policies Encouraging the Production of Agrofuels on the Enjoyment of the Human Rights to Food, Work and the Environment among the Peasant and Indigenous Communities and Rural Workers in Brazil*, FoodFirst International Network (FIAN), www.fian.org

FoE (2006) *Dirty Truths: Incineration and Climate Change*, London: Friends of the Earth (FoE).

FoE (2008) *Private Finance Initiative (PFI) Funding for Waste Infrastructure*, London: Friends of the Earth (FoE), www.foe.co.uk/sites/default/files/downloads/waste_pfi.pdf

FoEE (2010a) *The EU Emissions Trading System: Failing to Deliver*, Brussels: Friends of the Earth Europe (FoEE).

FoEE (2010b) *'Sustainable' Palm Oil Driving Deforestation: Biofuel Crops, Indirect Land Use Change and Emissions*, Brussels: Friends of the Earth Europe (FoEE).

FoEE (2012) *EU Biofuel Targets Will Cost €126 Billion Without Reducing Emissions*, Brussels: Friends of the Earth Europe (FoEE), www.foeeurope.org/EU-biofuel_cost-020212

FoEE (2014) *Agroecology: Building a New Food System for Europe*, Brussels: Friends of the Earth Europe (FoEE).

FoEE (2015) *Climate Myth #7: Technology Will Save Us*, Brussels: Friends of the Earth Europe (FoEE), www.foeeurope.org/yfoee/climate-mythbuster-technology-will-save-us

FoEE (2018) *Sufficiency: Moving Beyond the Gospel of Eco-Efficiency*, Brussels: Friends of the Earth Europe (FoEE), www.foeeurope.org/sites/defa ult/files/resource_use/2018/foee_sufficiency_booklet.pdf, https://friendsoft heearth.eu/news/sufficiency-a-call-for-transformative-and-systemic-change/

FoEE et al (2008) EU biofuels target must go: NGOs call for biofuels target to be dropped as UK review calls for an immediate rethink and MEPs go to vote, BirdLife International, European Environmental Bureau, Friends of the Earth Europe and Greenpeace, 7 July, www.greenpeace.org/eu-unit/ en/News/2009-and-earlier/biofuels-gallagher-report-07-07-08/

Franco, J., Levidow, L., Fig, D., Goldfarb, L., Hönicke, M. and Mendonça, M.L. (2010) Assumptions in the European Union biofuels policy: Frictions with experiences in Germany, Brazil and Mozambique, *Journal of Peasant Studies* 37(4): 661–698.

Freeman, C. (1991) Techno-economic paradigm and biological analogies in economics, *Revue économique* 42(2): 211–231.

Fressoli, M., Elisa Arond, Dinesh Abrol, Adrian Smith, Adrian Ely and Rafael Dias (2014) When grassroots innovation movements encounter mainstream institutions: Implications for models of inclusive innovation, *Innovation and Development* 4(2): 277–292.

Frickel, S., Gibbon, S., Howard, J., Kempner, J, Ottinger, G. and Hess, D. (2010) Undone science: Social movement challenges to dominant scientific practice, *Science, Technology, and Human Values* 35(4): 444–473.

GACSA (2014) *Framework Document, Global Alliance for Climate-Smart Agriculture* (GACSA), Rome: Food and Agriculture Organisation, www. fao.org/3/a-au667e.pdf

GAIA (2006) *Incinerators in Disguise: Case Studies of Gasification, Pyrolysis, and Plasma in Europe, Asia, and the United States*, Global Alliance for Incinerator Alternatives (GAIA), www.no-burn.org/wp-content/uploads/Incinerat ors-in-Disguise-Case-Studies-of-Gasification-Pyrolysis-and-Plasma-in-Eur ope-Asia-and-the-United-States-.pdf

GALBA (2021) *A Green New Deal For Leeds City Region: GALBA's Vision for a Sustainable Local Economy*, Leeds: Group for Action on Leeds-Bradford Airport (GALBA), www.galba.uk/green-new-deal

Galli, F. and Brunori, G. (eds) (2013) *Short Food Supply Chains as Drivers of Sustainable Development: Evidence Document*, FP7 project Foodlinks (GA No 265287), Laboratorio di studi rurali Sismondi.

GBC (2020) *The Retrofit Playbook: Driving Retrofit of Existing Homes – a Resource for Local and Combined Authorities*, London: UK Green Building Council (GBC), www.ukgbc.org/ukgbc-work/driving-retrofit-of-existing-homes/

GCA (2019) *The Contribution of Agroecological Approaches to Realizing Climate-Resilient Agriculture*, Global Commission on Adaptation (GCA), CGIAR Research Program on Forests Trees and Agroforestry (FTA), https://gca.org/wp-content/uploads/2020/12/TheContributionsOfAgroecologicalApproaches.pdf

Geels, F. (2014) Regime resistance against low-carbon transitions: introducing politics and power into the Multi-Level Perspective, *Theory, Culture & Society* 31(5): 21–40.

Gelderloos, P. (2022) *The Solutions are Already Here: Strategies for Ecological Revolution from Below*, London: Pluto.

GIB (2014) *The UK Residual Waste Market: A Market Report by the UK Green Investment Bank*, London: Green Investment Bank (GIB).

Glasgow City Council (2021) *Glasgow Green Deal: Our Roadmap and Call for Ideas.*

Glasgow City Region (2021a) *Home Energy Retrofit Programme*, https://invest-glasgow.foleon.com/igpubs/glasgow-greenprint-for-investment/glasgow-city-region-home-energy-retrofit-programme/

Glasgow City Region (2021b) *Home Energy Retrofit Final Report: Next Steps*, Home Energy Retrofit Programme, https://invest-glasgow.foleon.com/igpubs/glasgow-greenprint-for-investment/glasgow-city-region-home-energy-retrofit-programme/

GL Law (2019) Waste of money: Campaign to investigate the rising cost of incinerator reaches critical stage, *GL Law*, 22 March, www.gl.law/insight/news/procurement-dispute-incinerator/

GNDE (2019) *Blueprint for Europe's Just Transition: The Green New Deal for Europe*, Green New Deal for Europe (GNDE), www.gndforeurope.com

Goldberg, M.L. (2019) How a dispute over carbon market regulations tripped up COP25 in Madrid, *UN Despatch*, 17 December, www.undispatch.com/how-a-dispute-over-carbon-market-regulations-tripped-up-cop25-in-madrid/

Gottweis, H. (1998) *Governing Molecules: The Discursive Politics of Genetic Engineering in Europe and the United States*, Cambridge, MA: MIT Press.

Gouldson, A., Sudmant, A., Duncan, A. and Wiliamson, R. (2020) *A Net-Zero Carbon Roadmap for Leeds*, Leeds Climate Commssion & Place-Based Climate Action Network, www.leedsclimate.org.uk/sites/default/files/Net-Zero%20Carbon%20Roadmp%20for%20Leeds_0.pdf

GRAIN (2016) *Together We Can Cool the Planet*, Barcelona: GRAIN, https://grain.org/en/article/5620-comic-book-together-we-can-cool-the-planet

GRAIN and WRM (2015) *How REDD+ Projects Undermine Peasant Farming and Real Solutions to Climate Change*, Grain and World Rainforest Movement (WRM).

Green Car Congress (2006) 75% of major European car brands not tracking to meet voluntary CO2 reduction commitments, *Green Car Congress*, www.greencarcongress.com/2006/10/75_of_major_eur.html

Greenpeace and Runnymede Trust (2022) *Confronting Injustice: Racism and the Environmental Emergency*, London: Greenpeace & Runnymede Trust, www.greenpeace.org.uk/challenges/environmental-justice/race-enviro nmental-emergency-report/

Greenpeace USA (2015) *Carbon Capture Scam: How a False Climate Solution Bolsters Big Oil*, Washington, DC: Greenpeace USA.

Greens/EFA (2020) *The Myth of Climate Smart Agriculture: Why Less Bad isn't Good*, Brussels: Greens/EFA, www.martin-haeusling.eu/presse-med ien/publikation-in-englisch.html, www.arc2020.eu/the-myth-of-clim ate-smart-agriculture-why-less-bad-isnt-good/

Grubler, A., Wilson, C., Bento, N., Boza-Kiss, B., Krey, V., McCollum, D., et al (2018) A low energy demand scenario for meeting the 1.5 °C target and sustainable development goals without negative emission technologies, *Nature Energy* 3(6): 517–525.

GTM (2018) With 43 carbon-capture projects lined up worldwide, supporters cheer industry momentum, *Green Tech Media* (GTM), 11 December, www.greentechmedia.com/articles/read/carbon-capture-gains-momentum https://cles.org.uk/blog/climate-emergency-requi res-local-economic-restructuring/

Gupta, A.K. (1996) The Honey Bee Network: Voices from grassroots innovators, *Cultural Survival*, March, www.culturalsurvival.org/publications/ cultural-survival-quarterly/honey-bee-network-voices-grassroots-innovators

Guterres, A. (2022a) Press conference launch of IPCC report, United Nations, 28 February, https://media.un.org/en/asset/k1x/k1xcijxjhp

Guterres, A. (2022b) Secretary-General warns of climate emergency, United Nations, 4 April, www.un.org/press/en/2022/sgsm21228.doc.htm

Hadden, J. (2015) *Networks in Contention: The Divisive Politics of Climate Change*, Cambridge: Cambridge University Press.

Hajer, M.A. (1995) *The Politics of Environmental Discourse: Ecological Modernization and the Policy Process*, Oxford: Oxford University Press.

Hall, R. and Zacune, J. (2012) *Bio-economies: The EU's Real 'Green Economy' Agenda?* London: World Development Movement; Amsterdam: Transnational Institute.

Hansard (2020) Debate on waste incineration, *Hansard*, 11 February, https:// hansard.parliament.uk/Commons/2020-02-11/debates/D1799344-3D26-4DF0-94C1-3AEB397AF375/WasteIncinerationFacilities

Harrabin, R. (2021) John Kerry: US climate envoy criticised for optimism on clean tech, *BBC News*, 17 May, www.bbc.co.uk/news/science-envi ronment-57135506

Harrison, P. (2010) Europe finds politics and biofuels don't mix, *Reuters*, 5 July.

Harvey, D. (2005) *A Brief History of Neoliberalism*, Oxford: Oxford University Press.

Harvey, D. (2018) *The Limits to Capital*, London: Verso.

Harvie, P. (2022) Letter from the Minister for Zero Carbon Buildings, Active Travel and Tenants' Rights to the Convener, 11 January, Local Government, Housing and Planning Committee, Scottish Parliament, www.parliament. scot/chamber-and-committees/committees/current-and-previous-com mittees/session-6-local-government-housing-and-planning/corresponde nce/2022/retrofitting-housing-for-net-zero-january-2022

Hayns-W, S. (2018) Chief Defra scientist warns more incineration could harm innovation, *Resource*, 2 February, https://resource.co/article/chief-defra-scientist-warns-more-incineration-could-harm-innovation-12382

Heller, C. (2002) From scientific risk to paysan *savoir-faire*: Peasant expertise in the French and global debate over GM crops, *Science as Culture* 11(1): 5–37.

Heron, J. and Reason, P. (2006) The practice of co-operative inquiry: Research with rather than on people, in P. Reason and H. Bradbury (eds) *Handbook of Action Research: The Concise Paperback Edition*, London: SAGE, pp 144–154.

Heron, J. and Reason, P. (2008) Extending epistemology within a co-operative inquiry, in P. Reason and H. Bradbury (eds) *SAGE Handbook of Action Research: Participative Inquiry and Practice* (2nd edn), London: Sage, pp 366–380, https://ikhsanaira.files.wordpress.com/2016/09/action-resea rch-participative-inquiry-and-practice-reasonbradburry.pdf

Hess, D. (2007) *Alternative Pathways in Science and Industry*, Cambridge, MA: MIT Press.

Hess, D. (2016) *Undone Science: Social Movements, Mobilized Publics, and Industrial Transitions*, Cambridge, MA: MIT Press.

Hickel, J. (2020) *Less is More: How Degrowth Will Save the World*, London: Windmill.

Hilmi, A. (2012) *La Transition Agricole: Une autre logique*, Le Réseau Plus et Mieux, www.moreandbetter.org

Hinchliffe, S., Levidow, L. and Oreszczyn, S. (2014) Engaging cooperative research, *Environment and Planning A* 46: 2080–2094.

Hines, C. (2000) *Localization: A Global Manifesto*, London: Earthscan, www. eurovia.org/IMG/article_PDF_article_a302.pdf

HMG (2009) *The UK Renewable Energy Strategy*, London: Her Majesty's Government (HMG).

Huber, D. (2020) *The New European Commission's Green Deal and Geopolitical Language: A Critique from a Decentring Perspective*, Istituto Affari Internazionali (IAI), www.jstor.org/stable/resrep25276

Huber, M. (2022) *Climate Change as Class War: Building Socialism on a Warming Planet*, London: Verso.

Huesemann, M. and Huesemann, J. (2011) *Techno-Fix: Why Technology Won't Save Us or the Environment*, Philadelphia: New Society, http://newt echnologyandsociety.org/

Hultman, J. and Corvellec, H. (2012) The European waste hierarchy: From the sociomateriality of waste to a politics of consumption, *Environment and Planning A* 44: 2413–2427.

IAASTD (2009) *Agriculture at a Crossroads: Synthesis Report*, International Assessment of Agricultural Science, Technology and Development (IAASTD), www.agassessment.org/reports/iaastd/en/agriculture%20 at%20a%20crossroads_synthesis%20report%20(english).pdf

ICHRP (2008) *Climate Change and Human Rights: A Rough Guide*, Geneva: International Coalition for Human Rights in the Philippines (ICHRP).

IEA (2017) *Tracking Clean Energy Progress (TCEP) 2017*, Vienna: International Energy Agency (IEA), www.iea.org/reports/tracking-clean-energy-progr ess-2017

IEA (2019) *Global Energy & CO2 Status Report 2019*, Paris: International Energy Authority (IEA), www.iea.org/reports/global-energy-co2-sta tus-report-2019

IEA (2021a) *World Energy Balances: Statistics Report*, Paris: International Energy Agency (IEA).

IEA (2021b) *World Energy Investment 2021*, Paris: International Energy Agency (IEA), www.iea.org/reports/world-energy-investment-2021/ executive-summary

IEA (2021c) *The Role of Critical Minerals in Clean Energy Transitions*, Vienna: International Energy Agency (IEA).

IEA (2021d) *CCUS in Industry and Transformation*, Vienna: International Energy Agency (IEA), www.iea.org/reports/ccus-in-industry-and-tra nsformation

IEA (2022) Surging electricity demand is putting power systems under strain around the world, *International Energy Agency* (IEA), www.iea.org/news/ surging-electricity-demand-is-putting-power-systems-under-strain-aro und-the-world

IFOAM EU Group (2013) CAP deal lacks strong and credible steps towards sustainability, *IFOAM EU Group*, 26 June, http://eu.ifoam.org/en/news/ 2013/06/26/press-release-cap-deal-lacks-strong-and-credible-steps-towa rds-sustainability

IFOAM EU Group, ARC2020 and TP Organics (2012) *Agroecology: Ten Examples of Successful Innovation in Agriculture*, http://agro-ecoinnovation.eu/ wp-content/uploads/2012/11/Eco_Innovation_broch_24pages_ENG_lr.pdf

IFPRI (2010) *Global Trade and Environmental Impact Study of the EU Biofuels Mandate*, Washington, DC: International Food Policy and Research Institute (IFPRI).

Ilieva, R.T. and Hernandez, A. (2018) Scaling-up sustainable development initiatives: A comparative case study of agri-food system innovations in Brazil, New York, and Senegal, *Sustainability* 10: 4057, www.mdpi.com/ 2071-1050/10/11/4057.

IPES Food (2019) *Towards A Common Food Policy for the European Union*, International Panel of Experts on Sustainable Food Systems (IPES Food), http://agroecology-europe.org/wp-content/uploads/2019/02/CFP_FullReport.pdf

IPPR (2020) *All Hands to the Pump: A Home Improvement Plan for England*, London: Institute for Public Policy Research (IPPR).

Jacobson, M.Z. (2020) Evaluation of coal and natural gas with carbon capture as proposed solutions to global warming, air pollution, and energy security, in *100% Clean, Renewable Energy and Storage for Everything*, Cambridge: Cambridge University Press, https://web.stanford.edu/group/efmh/jacobson/WWSBook/WWSBook.html

Jasanoff, S. (1997) Civilization and madness: The great BSE scare of 1996, *Public Understanding of Science* 6(3): 221–232.

Jasanoff, S. (2004) *States of Knowledge: The Co-Production of Science and the Social Order*, New York: Routledge.

Jasanoff, S. (2015) Imagined and invented worlds, in S. Jasanoff and S.H. Kim (eds) *Dreamscapes of Modernity: Sociotechnical Imaginaries and the Fabrication of Power*, Chicago: University of Chicago Press, pp 321–342.

Jasanoff, S. and Kim, S.H. (2009) Containing the atom: Sociotechnical imaginaries and nuclear power in the United States and South Korea, *Minerva* 47(2): 119–146.

Jasanoff, S. and Kim, S.H. (eds) (2015) *Dreamscapes of Modernity: Sociotechnical Imaginaries and the Fabrication of Power*. Chicago: The University of Chicago Press.

Jessop, B. (2005) Cultural political economy, the knowledge-based economy, and the state, in A. Barry and D. Slater (eds) *The Technological Economy*, London: Routledge, pp 144–166.

Jessop, B. (2010) Cultural political economy and critical policy studies, *Critical Policy Studies* 3(3–4): 336–356.

Johnson, B. and Andersen, A.D. (2012) *Learning, Innovation and Inclusive Development*, Aalborg: Aalborg University Press.

Johnston, S.F. (2018) Alvin Weinberg and the promotion of the technological fix, *Technology and Culture* 59(3): 620–651.

Jordan, S. and Kapoor, D. (2016) Re-politicizing participatory action research: Unmasking neoliberalism and the illusions of participation, *Educational Action Research* 24(1): 134–149.

Josette, S. (2022) Community organising: A crucible for local, national and global action, *Municipal Enquiry*, www.municipal-enquiry.org/post/community-organising-a-crucible-for-local-national-and-global-action

JRC (2008) *Biofuels in the European Context: Facts and Uncertainties*, Ispra: Joint Research Centre (JRC), European Commission, JRC 43285.

Kaiser, W. and Schot, J. (2014) *Making the Rules for Europe: Experts, Cartels, International Organizations*, London: Palgrave.

Kaminski, I. (2015) Slow road to recovery, *Mineral and Waste Planning*, 2 April, www.mineralandwasteplanning.co.uk/slow-road-recovery/article/1341429

Karliner, J. (2000) Climate Justice Summit provides alternative vision, *Corporate Watch*, www.corpwatch.org/article/climate-justice-summit-provi des-alternative-vision

Karner, S. (ed) (2010) *Local Food Systems in Europe: Case Studies from Five Countries and What They Imply for Policy and Practice*, Graz: Inter-University Research Centre on Technology, Work and Culture (IFZ), Austria, www. genewatch.org/uploads/f03c6d66a9b354535738483c1c3d49e4/FAAN_ Booklet_PRINT.pdf

Kneafsey, M., Venn, L., Schmutz, U., Balázs, B., Trenchard, L., Eyden-Wood, P., Bos, E., Sutton, G. and Blackett, M. (2013) *Short Food Supply Chains and Local Food Systems in the EU: A State of Play of Their Socio-economic Characteristics*. Brussels, European Commission, http://dx.doi. org/10.2791/88784

Kinchy, A. (2012) *Seeds, Science, and Struggle: The Global Politics of Transgenic Crops*, Cambridge, MA: MIT Press.

Klein, N. (2014) *This Changes Everything: Capitalism vs. the Climate*, New York: Simon & Schuster.

Klein, N. (2019) *On Fire: The (Burning) Case for a Green New Deal*, New York: Simon & Schuster.

Klimaforum09 (2009) *System Change – not Climate Change: A People's Declaration from Klimaforum09*, http://klimaforum.org/declaration_english.pdf

Kreuter, J. (2015) *Technofix, Plan B or Ultima Ratio? A Review of the Social Science Literature on Climate Engineering Technologies*, University of Oxford, Institute for Science, Innovation and Society.

Krüger, T. (2017) Conflicts over carbon capture and storage in international climate governance, *Energy Policy* 100(1): 58–67.

Krukowska, E. (2021) COP26 finally set rules on carbon markets: What does it mean?, *Bloomberg*, 13 November.

Kurtzleben, D. (2021) Ocasio-Cortez sees Green New Deal progress in Biden Plan, but 'it's not enough', *NPR*, 5 April, www.npr.org/2021/04/ 02/983398361/green-new-deal-leaders-see-biden-climate-plans-as-a-vict ory-kind-of?t=1658736251428

LabGND (2019a) Green New Deal explained, *Labour for a Green New Deal* (LabGND), www.labourgnd.uk/gnd-explained

LabGND (2019b) Labour commits to decarbonisation by 2030 with Green New Deal, *Labour for a Green New Deal* (LabGND), 24 September, www. labourgnd.uk/news/2019/9/24/labour-backs-gnd

Labour Party (2019a) *Labour Together Submission*, Labour Party Community Organising Unit, https://uploads-ssl.webflow.com/5e68e49babaae714a a0b77b2/5ee8dd734cf74dc531c82247_Community%20Organising%20U nit%20Submission.pdf

Labour Party (2019b) *It's Time for Real Change: Labour Party Manifesto 2019*, London: Labour Party, https://labour.org.uk/manifesto-2019/a-green-ind ustrial-revolution

Labour Party (2019c) *Labour's Green-Transformation-Fund*, London: Labour Party, https://labour.org.uk/manifesto-2019/a-green-industrial-revolution/

Lakner, S. (2016) Lighter shade of green: CAP fails in Germany & beyond, https://slakner.wordpress.com/2016/04/13/greening-of-direct-payme nts-first-preliminary-figures-on-the-eu-level

Laurance, W.F. (2007) Switch to corn promotes Amazon deforestion, *Science* 318(5857): 1721.

Law, J. (ed) (1986) *Power, Action and Belief: A New Sociology of Knowledge*, London: Routledge.

LCH (2021) Pre-COP26, we revisit Glasgow's retrofit scene, *Low Carbon Homes* (LCH), 24 May, www.lowcarbonhomes.uk/news/retrofit-revisi ted-glasgow/

Lecain, T. (2004) When everybody wins does the environment lose? The environmental techno-fix in twentieth century American mining, in L. Rosner (ed) *The Technological Fix: How People Use Technology to Create and Solve Social Problems*, New York: Routledge, pp 137–151.

Leeds City Council (2021) Council housing project claims top regional green award, *Leeds City Council*, 19 May, https://news.leeds.gov.uk/news/ council-housing-project-claims-top-regional-green-award

Leeds Climate Commission (2019) *Hydrogen Conversion: Potential Contribution to a Low Carbon Future for Leeds*, www.leedsclimate.org.uk/hydrogen-con version-potential-contribution-low-carbon-future-leeds

Leeds Trades Council (2020) Retrofit Leeds homes with high-quality insulation and heat pumps: a plan and call to action!, *Leeds Trades Council*, https://leedstuc.files.wordpress.com/2020/09/draft-document-decarbonis ing-leeds-homes-with-a-huge-programme-of-deep-retrofitting-and-insta llation-of-heat-pumps.pdf

Lehtonen, M. (2011) Social sustainability of the Brazilian bioethanol: Power relations in a centre-periphery perspective, *Biomass and Bioenergy* 35: 2425–2434.

Let's Recycle (2004) RDF should be landfilled, says Friends of the Earth, *Let's Recycle*, 22 April, www.letsrecycle.com/news/latest-news/rdf-sho uld-be-landfilled-says-friends-of-the-earth/

Levidow, L. (2000) Pollution metaphors in the UK biotechnology controversy, *Science as Culture* 9(3): 325–351.

Levidow, L. and Bijman, J. (2002) Farm inputs under pressure from the European food industry, *Food Policy* 27(1): 31–45.

Levidow, L. and Paul, H. (2011) Global agrofuel crops as contested sustainability, Part II: Eco-efficient techno-fixes?, *Capitalism Nature Socialism* 22(2): 27–51.

Levidow, L. and Neubauer, C. (2012) Opening up societal futures through EU research and innovation agendas, *EASST Review*, 31(3): 4–11, http://easst.net/wp-content/uploads/2012/09/review_2012_09.pdf

Levidow, L. and Oreszczyn, S. (2012) Challenging unsustainable development through research cooperation, *Local Environment: The International Journal of Justice and Sustainability*, 17(1): 35–56.

Levidow, L. and Papaioannou, T. (2016) Policy-driven, narrative-based evidence-gathering: UK priorities for decarbonisation through biomass, *Science and Public Policy* 43(1): 46–61.

Levidow, L. and Papaioannou, T. (2018) Which inclusive innovation? Competing normative assumptions around social justice, *Innovation & Development* 8(2): 209–226.

Levidow, L., Birch, K. and Papaioannou.,T. (2013) Divergent paradigms of European agro-food innovation: The Knowledge-Based Bio-Economy (KBBE) as an R&D agenda, *Science, Technology and Human Values* 38(1): 94–125.

Levidow, L., Carr, S. and Wield, D. (2000) Genetically modified crops in the European Union: Regulatory conflicts as precautionary opportunities, *Journal of Risk Research* 23(3): 189–208.

Levidow, L., Carr, S. and Wield, D. (2005) EU regulation of agri-biotechnology: Precautionary links between science, expertise and policy, *Science & Public Policy* 32 (4): 261–276.

Levidow, L., Carr, S., von Schomberg, R. and Wield, D. (1996) Regulating agricultural biotechnology in Europe: Harmonization difficulties, opportunities, dilemmas, *Science & Public Policy* 23(3): 135–157.

Levidow, L., Søgaard, V. and Carr, S. (2002) Agricultural PSREs in Western Europe: Research priorities in conflict, *Science and Public Policy* 29(4): 287–295.

Levy, D.L. and Spicer, A. (2013) Contested imaginaries and the cultural political economy of climate change, *Organization* 20(5): 659–678.

Lewin, K. (1946) Action research and minority problems, in G.W. Lewin (ed) *Resolving Social Conflict*, London: Harper & Row, pp 143–152.

LGHPC (2021) *Retrofitting Housing For Net Zero*, Local Government, Housing and Planning Committee (LGHPC), 30 November, Edinburgh: Scottish Parliament.

Lipietz, A. (2008) Food or fuel?. *Research Review*, May [European Parliament magazine], http://lipietz.net/Food-or-fuel

Lohmann, L. (2009) Climate as investment, *Development and Change* 40(6): 1063–1083.

Lohmann, L. (2020) Carbon markets do not need to be 'fixed': They need to be eliminated, *REDD Monitor*, https://redd-monitor.org/2020/10/22/interview-with-larry-lohmann-the-corner-house-carbon-markets-do-not-need-to-be-fixed-they-need-to-be-eliminated/

Lohmann, L. and Hildyard, N. (2013) *Energy Alternatives: Surveying the Territory*, The Corner House, www.thecornerhouse.org.uk/sites/thecorn erhouse.org.uk/files/ENERGY%20ALTERNATIVES%20--%20SU RVEYING%20THE%20TERRITORY.pdf

Lund Declaration (2009) *Europe Must Focus on the Grand Challenges of Our Time*, www.se2009.eu/polopoly_fs/1.8460!menu/standard/file/lund_decla ration_final_version_9_july.pdf

LVC (2009) *Feeding the World and Cooling the Planet*, La Vía Campesina's Fifth International Conference, La Via Campesina (LVC), https://viaca mpesina.org/en/feeding-the-world-and-cooling-the-planet-la-vcampesi nas-fifth-international-conference/

LVC (2013) *From Maputo to Jakarta: 5 Years of Agroecology in La Vía Campesina.* Jakarta: La Vía Campesina.

Magnien, E. and de Nettancourt, D. (1993) What drives European biotechnological research?, in E.J. Blakelely and K.W. Willoughby (eds) *Biotechnology Review no. 1: The Management and Economic Potential of Biotechnology*, Brussels: Commission of the European Communities, pp 47–48.

Malm, A. (2016) *Fossil Capital: The Rise of Steam Power and the Roots of Global Warming.* London: Verso.

Mandelson, P. (2007) *The Biofuel Challenge*, speech at Biofuels conference, 5 July, Brussels: DG Trade, European Commission.

Markusson, N., Gjefsen, M.D., Stephens, J.C. and Tyfield, D. (2017) The political economy of technical fixes: The (mis)alignment of clean fossil and political regimes, *Energy Research & Social Science* 23: 1–10.

Martinez-Alier, J. (2002) *The Environmentalism of the Poor: A Study of Ecological Conflicts and Valuation*, London: Edward Elgar.

Martínez-Torres, M.E. and Rosset, P.M. (2014) Diálogo de saberes in La Vía Campesina: Food sovereignty and agroecology, *Journal of Peasant Studies* 41(6): 979–997.

Marx, L. (1983) Are science and society going in the same direction?, *Science, Technology, & Human Values* 8(4): 6–9.

Mascarenhas, M. and Busch, L. (2006) Seeds of change: Intellectual property rights, genetically modified soybeans and seed saving in the United States, *Sociologia Ruralis* 46(2): 122–138.

Maslow, A. (1966) *The Psychology of Science: A Reconnaissance*, New York: Harper & Row.

Massot, A. (2019) *Towards a Post-2020 Common Agricultural Policy*, Brussels: European Parliament, www.europarl.europa.eu/factsheets/en/sheet/113/ towards-a-post-2020-common-agricultural-policy

Matondi, B., Havnevik, K. and Beyene, A. (2011) *Biofuels, Land Grabbing and Food Security in Africa*, London: Zed Books.

McInroy, N. (2020) Climate emergency requires local economic restructuring, *Local Government Chronicle*, 26 February, www.lgcplus.com/politics/governance-and-structure/climate-emergency-means-we-must-restructure-our-local-economies-17-02-2020/

McKechnie, S. (1999) Food fright, *The Guardian*, 10 February.

McLaren, D. and Markusson, N. (2020) The co-evolution of technological promises, modelling, policies and climate change targets, *Nature Climate Change* 10(5): 392–397, www.nature.com/articles/s41558-020-0740-1

McMichael, P. (2005) Global development and the corporate food regime, *Research in Rural Sociology and Development* 11: 265–299.

Melrose, M.J. (2001) Maximizing the rigor of action research: Why would you want to? How could you?, *Field Methods* 13(2): 160–180.

Mendonça, M.L. (2006) *Agroenergy: Myths and Impacts in Latin America*, Organization Pastoral Land.

Mendonça, M.L. (2010) *Monopólio da Terra no Brasil: Impactos da expansão de monocultivos para a produção de agrocombustíveis*, São Paulo: Rede Social de Justiça e Direitos Humanos & Comissão Pastoral da Terra.

Merrell, A. (2019) Protesters fan flames of controversy around incinerator, *Punchline Gloucester*, 8 January, www.punchline-gloucester.com/articles/aanews/protesters-fan-flames-of-controversy-ubaser-balfour-beatty-glou cestershire-incinerator-county-counci

Michalopolous, S. (2018) European market 'open to palm oil', EU ambassador in Malaysia says, *Euractiv*, 9 August, www.euractiv.com/sect ion/agriculture-food/news/european-market-open-to-palm-oil-eu-amb assador-in-malaysia-says/

Miller, E. (2010) Solidarity economy: Key concepts and issues, in E. Kawano, T. Masterson and J. Teller-Ellsberg (eds) *Solidarity Economy I: Building Alternatives for People and Planet*, Amherst: Center for Popular Economics, www.communityeconomies.org/sites/default/files/paper_attachment/Miller_Solidarity_Economy_Key_Issues_2010.pdf

Mol, A.P.J (1996) Ecological modernisation and institutional reflexivity: Environmental reform in the late modern age, *Environmental Politics* 5(2): 302–323.

Monnet, J. (1974) *L'Europe et la nécessité*, Archives de la Fondation Jean Monnet pour l'Europe.

Monsanto (1997) *Report on Sustainable Development*, St Louis: Monsanto Company.

Morgan, S. (2018) Palm oil ban on the ropes as Commission weighs options, *Euractiv*, 20 November, www.euractiv.com/section/agriculture-food/news/palm-oil-ban-on-the-ropes-as-commission-weighs-options

Morozov, E. (2013) *To Save Everything, Click Here: The Folly of Technological Solutionism*. London: Penguin.

Morozov, E. (2020) The tech 'solutions' for coronavirus take the surveillance state to the next level, *The Guardian*, 15 April.

MPS (2007) Wight means green in the UK, *Modern Power Systems* (MPS), 1 July, www.modernpowersystems.com/features/featurewight-means-green-in-the-uk

MTE (2010) *Resultados das operações de fiscalização para erradicação do trabalho escravo e annual*, Brasilia: Ministerio do Trabalho e Emprego (MTE).

Neale, J. (2021) *Fight the Fire: Green New Deals and Global Climate Change*, London: Resistance Books, https://theecologist.org/fight-the-fire

Neujeffski, M. (2021) Regions and the European Green Deal, *RegioParl*, 15 December, www.regioparl.com/regions-and-the-european-green-deal/?lang=en

NGN (2018) *H21 North of England Report*, Leeds: Northern Gas Networks (NGN), www.h21.green/projects/downloads/

Niggli, U., Slabe, A., Schmid, O., Halberg, N. and Schluter, M. (2008) *Vision for an Organic Food and Farming Research Agenda to 2025*, Brussels: IFOAM-EU Group, www.organic-research.org/index.html, http://orgprints.org/13439/

North, P. (2010) Eco-localisation as a progressive response to peak oil and climate change: A sympathetic critique, *Geoforum* 41(4): 585–594.

North, P. and Cato, M.S. (eds) (2017) *Towards Just and Sustainable Economies: The Social and Solidarity Economy North and South*, Bristol: Policy Press.

North, P. and Longhurst, N. (2013) Grassroots localisation? The scalar potential of and the limits of the 'transition' approach to climate change and resource constraint, *Urban Studies* 50: 1423–1438.

North, P., Nowak, V., Southern, A. and Thompson, M. (2020) Generative anger: From social enterprise to antagonistic economies, *Rethinking Marxism* 32(3): 330–347.

Nyari, B. (2008) *Biofuel Land Grabbing in Northern Ghana*, Regional Advisory and Information Network Systems (RAINS).

Nyéléni (2007) *Declaration of the Forum for Food Sovereignty*, Nyéléni.

Nyéléni Europe and Central Asia (2021) *Roots of Resilience: Land Policy for an Agroecological Transition in Europe*, www.arc2020.eu/wp-content/uploads/2021/03/rootsofresilience_online_final_en.pdf

OECD (2011) *Towards Green Growth: A Summary for Policy Makers*, Paris: OECD.

Ofgem (2015) Renewables Obligation Certificate (ROC) and Renewable Levy Exemption Certificate (LEC) Issue Schedule 2015–2016.

Okereke, C. (2007) *Global Justice and Neoliberal Environmental Governance: Ethics, Sustainable Development and International Co-operation*, London: Routledge.

OPSAL (2021) *Salares Andinos: Ecologia de Saberes por la Proteccion de Nuestros Salares y Humedales*, Observatorio Pluralnacional de Salares Andinos (OPSAL), Fundación Tantí.

Our Future Leeds (2021) Campaigns, www.ourfutureleeds.org/campaigns

PAC (2021) *Green Homes Grant Voucher Scheme*, London: UK Parliament, Public Accounts Committee (PAC), https://committees.parliament.uk/committee/127/public-accounts-committee/news/159264/pac-report-green-homes-grant-scheme-underperformed-badly/

Paciaroni, S. (2021) Springfield Cross: Low carbon social housing project takes shape Glasgow's East End, *Glasgow Times*, 8 November, www.glasgowtimes.co.uk/news/19701701.springfield-cross-low-carbon-social-housing-project-takes-shape-glasgows-east-end/

Page, E.A. (2012) The hidden costs of carbon commodification: Emissions trading, political legitimacy and procedural justice, *Democratization* 19(5): 932–950.

Parfitt, C. (2013) How are genetic enclosures shaping the future of the agrifood sector?, *New Zealand Sociology* 28(4): 33–58.

Paul, H. (2013) *A Foreseeable Disaster: The European Union's Agroenergy Policies and the Global Land and Water Grab*, Amsterdam: Transnational Institute.

Paul, H. (2015) Climate-smart agriculture: preparing for a corporate soil and climate-grab in Paris?, *The Ecologist*, 26 November, www.theecologist.org/essays/2986268/climatesmart_agriculture__preparing_for_a_corporate_soil_and_climategrab_in_paris.html

PCAN (2020) Accelerating retrofit webinar, *Place-Based Climate Action Network* (PCAN), 11 August, https://pcancities.org.uk/event/accelerating-retrofit

Peake, L. (2016) Advanced conversion technologies: A heated debate, *Resource*, 23 November, http://resource.co/article/advanced-conversion-technologies-heated-debate-11503

Peake, L. (2020) Scandinavians call their waste incineration 'crazy', so why copy them?, *Green Alliance Blog*, https://greenallianceblog.org.uk/tag/policy-connect/

Pe'er, G., Dicks, L.V., Visconti, P., Arlettaz, R., Báldi, A, Benton, T.G., et al. (2014) EU agricultural reform fails on biodiversity, *Science* 344(6188): 1090–1092, doi:10.1126/science.1252254

Perchard, E. (2015) Gloucestershire incinerator opponents submit alternative, *Resource*, 10 July, https://resource.co/article/gloucestershire-incinerator-opponents-submit-alternative-10278

Perez, C. (2009) Technological revolutions and techno-economic paradigms, *Cambridge Journal of Economics*, 34(1): 185–202.

Pielke, P. (2010) *The Climate Fix: What Scientists and Politicians Won't Tell You About Global Warming*, New York: Basic Books.

Pimbert, M. (2017) Agroecology as an alternative vision to conventional development and climate-smart agriculture, *Development* 58(2–3): 286–298, https://dx.doi.org/10.1057/s41301-016-0013-5

Planning Inspectorate (2015) Appeal Decision, Appeal Ref: APP/H4315/A/14/2224529, Former Ravenhead Glass Warehouse and other land, Lock Street, St Helens, WA9 1HS, www.nottinghamshire.gov.uk/media/110 190/document-ip12-supplementary-representation-from-ukwin-dated-13th-august-2015-appendix.pdf

Policy Connect (2020) *No Time to Waste: Resources, Recovery & the Road to Net-Zero*, All-Party Parliamentary Sustainable Resource Group and Sustainable Resource Forum, www.policyconnect.org.uk/research/no-time -waste-resources-recovery-road-net-zero

Rainforest Alliance (2020) What's in Our 2020 Certification Program? Climate-Smart Agriculture, www.rainforest-alliance.org/resource-item/whats-in-our-2020-certification-program-climate-smart-agriculture/

Rankin, J. (2011) Growing pressure to change EU biofuel policy, *European Voice*, 28 April, www.europeanvoice.com/article/imported/growing-press ure-to-change-eu-biofuel-policy/70930.aspx

REA (2011) *EfW: Guide for Decision Makers*. London: Renewable Energy Association.

REA (2014) *Gasification and Pyrolysis*. London: Renewable Energy Association.

Reece, A. (2013) Alternatives to Javelin Park incinerator, *Resource*, 17 May, https://resource.co/article/Latest/Alternatives_Javelin_Park_incinera tor-3097

Reitan, R. and Gibson, S. (2012) Climate change or social change? Environmental and leftist praxis and participatory action research, *Globalizations* 9(3): 395–410.

Reno, J. (2011) Motivated markets: Instruments and ideologies of clean energy in the United Kingdom, *Cultural Anthropology* 26(3): 389–413.

Reyes, O. (2012) What goes up must come down: carbon trading, industrial subsidies and capital market governance, in N. Hallström (ed) *What Next: Climate, Development and Equity*, Uppsala: Dag Hammarskjöld Foundation, pp 185–209, www.daghammarskjold.se/wp-content/uplo ads/2012/09/Climate-Development-and-Equity_single_pages.pdf

Reyes, O. (2014) *Life Beyond Emissions Trading*, Brussels: Corporate Europe Observatory (CEO).

Ribeiro, D. and Matavel, N. (2009) *Jatropha! A Socio-economic Pitfall for Mozambique*, Maputo: Justiça Ambiental & União Nacional de Camponeses (UNAC), www.swissaid.ch/global/PDF/entwicklungspolitik/agrotreibsto ffe/Report_Jatropha_JA_and_UNAC.pdf

Ribeiro-Broomhead, J. and Tangri, N. (2021) *Zero Waste and Economic Recovery: The Job Creation Potential of Zero Waste Solutions*, Global Alliance for Incinerator Alternatives (GAIA), www.doi.org/10.46556/GFWE6885

Ridley, D. (2020) Laboratories for green future, *Red Pepper*, Summer: 20–22, www.redpepper.org.uk/

Riofrancos, T. (2019) Plan, mood, battlefield – reflections on the Green New Deal, *Viewpoint Magazine*, viewpointmag.com/2019/05/16/plan-mood-battlefield-reflections-on-the-green-new-deal/

Riofrancos, T. (2020) *Resource Radicals: From Petro-Nationalism to Post-Extractivism in Ecuador*. Durham, NC: Duke University Press.

Riofrancos, T. (2022) Shifting mining from the global South misses the point of Climate Justice, *Foreign Policy*, https://foreignpolicy.com/2022/02/07/renewable-energy-transition-critical-minerals-mining-onshoring-lithium-evs-climate-justice/

RIPESS (2011) *RIPESS Charter*, European Network of the Social Solidarity Economy, Réseau Intercontinental de Promotion de l'Économie Sociale Solidaire (RIPESS), https://ripess.eu/en/about-us/ripess-charter

RIPESS (2015) *Global Vision for a Social Solidarity Economy: Convergences and Differences in Concepts, Definitions and Frameworks*, European Network of the Social Solidarity Economy, Réseau Intercontinental de Promotion de l'Économie Sociale Solidaire (RIPESS), www.ripess.org/wp-content/uploads/2017/08/RIPESS_Vision-Global_EN.pdf

Rollinson, A.N. (2019) Efficiency and performance assessment of waste-to-energy melting gasification in relation to the EU waste framework directive, *Waste Management Volume 9, Waste to Energy*. Neuruppin: TK Verlag Karl Thomé-Kozmiensky, pp 371–382.

Rootes, C. (2009) More acted upon than acting? Campaigns against waste incinerators in England, *Environmental Politics* 18(6): 869–895.

Rosamond, B. (2002) Imagining the European economy: 'Competitiveness' and the social construction of Europe as an economic space, *New Political Economy* 7(2): 157–177.

Rosner, L. (ed) (2004) *The Technological Fix: How People Use Technology to Create and Solve Social Problems*, New York: Routledge.

Rosset, P.M. and Martínez-Torres, M.E. (2012) Rural social movements and agroecology: context, theory, and process, *Ecology and Society* 17(3): 17, http://dx.doi.org/10.5751/ES-05000-170317

Rowlatt, J. (2020) Greta Thunberg: Climate change 'as urgent' as coronavirus, *BBC News*, 20 June, www.bbc.co.uk/news/science-environment-53100800

Roy, I. (2020) Waste incinerators three times more likely to be sited in UK's most deprived neighbourhoods, *Unearthed*, https://unearthed.greenpeace.org/2020/07/31/waste-incinerators-deprivation-map-recycling/

RWM (2014) *Ever Decreasing Circles: Closing In on the Circular Economy*, RWM Ambassadors in collaboration with CIWM.

Sandlands, D. (2021) 'Massive' energy retrofit programme could target over 420,000 homes across Glasgow, *Glasgow Live*, 11 April, www.glasgowlive.co.uk/news/glasgow-news/glasgow-energy-refit-housing-programme-20360138

Sarewitz, D. (1996) *Frontiers of Illusion: Science, Technology and Politics of Progress.* Philadelphia: Temple University Press.

SCAR FEG (2008) *2nd Foresight Exercise: New Challenges for Agricultural Research: Climate Change, Food Security, Rural Development, Agricultural Knowledge Systems,* Brussels: Standing Committee on Agricultural Research (SCAR), Foresight Expert Group (FEG), http://ec.europa.eu/research/agriculture/scar/foresight_en.htm

SCAR FEG (2011) *Sustainable Food Consumption and Production in a Resource-Constrained World.* Brussels: Standing Committee on Agricultural Research, Foresight Expert Group. http://ec.europa.eu/research/agriculture/scar/pdf/scar_feg3_final_report_01_02_2011.pdf

Schellnhuber, H.J. (2011) Geoengineering: The good, the MAD, and the sensible, *PNAS* 108(51): 20277–20278.

Schimel, D., Stephens, B.B. and Fisher, J.B. (2015) Effect of increasing CO_2 on the terrestrial carbon cycle, *Proceedings of the National Academy of Sciences* 112: 436–441.

Schlissel, D. and Wamsted, D. (2018) *Holy Grail of Carbon Capture Continues to Elude Coal Industry,* Lakewood: Institute for Energy Economics and Financial Analysis.

Schlosberg, D. (2013) Theorising environmental justice: The expanding sphere of a discourse, *Environmental Politics* 22(1): 37–55.

Schmid, O., Padel, S. and Levidow, L. (2012) The bio-economy concept and knowledge base in a public goods and farmer perspective, *Bio-based and Applied Economics (BAE)* 1(1): 47–63, https://orgprints.org/id/eprint/20942/1/SCHMID_BAE_2012_10770-18316-1-PB.pdf

Schmid, O., Padel, S., Halberg, N., Huber, M., Darnhofer, I., Micheloni, C., et al (2009) *Strategic Research Agenda for Organic Food and Farming,* Brussels: Technology Platform Organics, www.organic-research.org/index.html

Scholes, P. (2014) The future of landfill tax, *Local Authority & Waste Recycling* (LAWR) 22(4): 11–12.

Schulze, E.-D. (2012) Large-scale bioenergy from additional harvest of forest biomass is neither sustainable nor greenhouse gas neutral, *GCB Bioenergy* 4(6): 611–616, https://onlinelibrary.wiley.com/doi/full/10.1111/j.1757-1707.2012.01169.x

Schurman, R. and Munro, W.A. (2010) *Fighting for the Future of Food: Activists versus Agribusiness in the Struggle over Biotechnology,* Minneapolis: University of Minnesota Press

Schweiger, T. (2001) Europe: hostile lands for GMOs, in B. Tokar (ed) *Redesigning Life? The Worldwide Challenge of Genetic Engineering,* London: Zed, pp 361–372.

Scot.E3 (2021) *Briefing #13: The Use & Abuse of Hydrogen,* https://scote3.files.wordpress.com/2021/12/briefing-13.pdf

Scott, N.D. (2011) The technological fix criticisms and the agricultural biotechnology debate, *Journal of Agricultural and. Environmental Ethics* 24: 207–226.

Searchinger, T. (2008) *The Impacts of Biofuels on Greenhouse Gases: How land use change alters the equation.* The German Marshall Fund of the United States, Economic Policy Program.

Searchinger, T., Heimlich, R., Houghton, R.A., Dong, F., Elobeid, A., Fabiosa, J., et al (2008) Use of U.S. cropland for biofuels increases greenhouse gases through emissions from land-use change, *Science* 319(5867): 1238–1240.

Seltenrich, N. (2013) Incineration versus recycling: In Europe, a debate over trash, *e360*, 28 August, http://e360.yale.edu/feature/incineration_versus_recycling__in_europe_a_debate_over_trash/2686

Sevilla Guzmán, E. and Woodgate, G. (2013) Agroecology: Foundations in agrarian social thought and sociological theory, *Agroecology and Sustainable Food Systems* 37(1): 32–44.

Sharman, A. (2009) *Evidence-based Policy or Policy-Based Evidence Gathering? The Case of the 10% Target*, MSc thesis, School of Geography and the Environment, University of Oxford.

Sharman, A. and Holmes, J. (2010) Evidence-based policy or policy-based evidence gathering? Biofuels, the EU and the 10% target, *Environmental Policy and Governance* 20: 309–321.

Sigg, A. (2014) 'Mass burn' is Advanced Conversion Technology [PPt file]. Alfred Sigg, Director of Operations. INOVA/Hitachi Zosen.

Šimunović, N., Hesser, F. and Stern, T. (2018) Frame analysis of ENGO conceptualization of sustainable forest management: Environmental justice and neoliberalism at the core of sustainability, *Sustainability* 10: 3165, www.mdpi.com/2071-1050/10/9/3165

Smith, A. and Stirling, A. (2016) *Grassroots Innovation and Innovation Democracy*, London: STEPs Centre.

Smith, A., Ely, A. and Jones, P. (2019) *Doing Digital Differently: Four Innovation Lessons from the Grassroots*, Sussex: STEPs Centre, https://steps-centre.org/blog/doing-digital-differently-four-innovation-lessons-from-the-grassroots/

Smith, A., Fressoli, M. and Thomas, H. (2014) Grassroots innovation movements: Challenges and contributions, *Journal of Cleaner Production* 63: 114–124.

Smith, A., Fressoli, M., Abrol, D. and Ely, A. (2016) *Grassroots Innovation Movements*, London: Routledge.

Smith, A., Hielscher, S. and Fressoli, M. (2015) *Transformative Social Innovation Narrative: FabLabs*, TRANSIT: EU SSH.2013.3.2-1, www.transitsocialinnovation.eu

Smith, J. (2010) *Biofuels and the Globalisation of Risk*, London: Zed.

Smolker, R. (2015) COP21's climate technofix: Spinning carbon into gold and the myth of 'negative emissions', *The Ecologist*, 3 December, https://theecologist.org/2015/dec/03/cop21s-climate-technofix-spinning-carbon-gold-and-myth-negative-emissions

Snow, D. and Benford, R. (2000) Framing processes and social movements: An overview and assessment, *Annual Review of Sociology* 26: 611–639.

Snow, D., Rochford, B., Worden, S. and Benford, R. (1986) Frame alignment processes, micromobilization, and movement participation, *American Sociological Review* 51(4): 464–481.

Source News (2019) *A Green New Deal for Scotland*, https://sourcenews.scot/a-green-new-deal-for-scotland/

Spice (2013) *Treatment Options for Residual Waste*, Edinburgh: Scottish Parliament Information Centre (Spice).

SRU (2007) *Climate Change Mitigation by Biomass*, Special Report, Sachverständigenrat für Umweltfragen (SRU), Advisory Council on the Environment, http://eeac.hscglab.nl/files/D-SRU_ClimateChangeBiomass_Jul07.pdf

Stuart, D. (2022) Tensions between individual and system change in the climate movement: An analysis of Extinction Rebellion, *New Political Economy* 27(5): 806–819.

STUC (2021) *Our Climate: Our Homes*, Scottish Trades Union Congress (STUC), https://stuc.org.uk/files/campaigns/Homes/Our-Homes_briefing.pdf

Sunrise Movement (2018) *What is the Green New Deal?*, www.sunrisemovement.org/green-new-deal/

Surin Declaration (2012) *First Global Encounter on Agroecology and Peasant Seeds*, http://viacampesina.org/en/index.php/main-issues-mainmenu-27/sustainable-peasants-agriculture-mainmenu-42/1334-surin-declaration-first-global-encounter-on-agroecology-and-peasant-seeds

Szabo, J. (2021) Fossil capitalism's lock-ins: The natural gas-hydrogen nexus, *Capitalism Nature Socialism* 32(4): 91–110.

Szerszynski, B., Kearnes, M., Macnaghten, P., Owen, R. and Stilgoe, J. (2013) Why solar radiation management geoengineering and democracy won't mix, *Environment and Planning* A 45(12): 2809–2816.

T&E (2006) Volkswagen failing on climate; Renault on track to meet EU target, 25 October, Brussels: Transport & Environment (T&E).

T&E (2008) NGOs call for suspension of biofuels targets, January, Brussels: Transport & Environment (T&E), www.transportenvironment.org/discover/ngos-call-suspension-biofuels-targets/

T&E (2009) Vans regulation carries fingerprints of strong industry lobbying, Brussels: Transport & Environment (T&E), www.transportenvironment.org/news/vans-regulation-carries-fingerprints-strong-industry-lobbying

T&E (2010a) T&E and other NGOs take legal action against Commission, *Bulletin*, March, Brussels: Transport & Environment (T&E), www.trans portenvironment.org/discover/te-and-other-ngos-take-legal-action-agai nst-commission

T&E (2010b) *Bioenergy: A Carbon Accounting Time Bomb*, Brussels: Transport and Environment (T&E), with Birdlife International and EEB.

T&E (2012a) Lobbying by car industry weakens 2020 action plan, Brussels: Transport & Environment (T&E), www.transportenvironment. org/news/lobbying-car-industry-weakens-2020-action-plan

T&E (2012b) Joint letter from ten NGOs to Commissioner Günther Oettinger, 25 September, Brussels: Transport & Environment (T&E).

T&E (2013) A weak cars CO2 deal better than no deal, Brussels: Transport & Environment (T&E), www.transportenvironment.org/news/%E2%80%98-weak-cars-co2-deal-better-no-deal%E2%80%99

T&E (2021) 10 years of EU's failed biofuels policy has wiped out forests the size of the Netherlands, Brussels: Transport & Environment (T&E), www. transportenvironment.org/discover/10-years-of-eus-failed-biofuels-pol icy-has-wiped-out-forests-the-size-of-the-netherlands-study

Tait, J., Chataway, J. and Wield, D. (2002) The life science industry sector: Evolution of agro-biotechnology in Europe, *Science and Public Policy* 29(4): 253–258.

Taylor, K. (2021) EU ministers attack plans to extend carbon pricing to heating and transport, *Euractiv*, 20 July.

Terrer, C. et al (2021) A trade-off between plant and soil carbon storage under elevated CO2, *Nature* 591: 599–603, www.nature.com/articles/s41 586-021-03306-8

Tighe, C. (2016) Teesside investor drops £300m renewables project, *Financial Times*, 5 April.

TNI (2015) *The Bioeconomy: A Primer*, Amsterdam: Transnational Institute (TNI), www.tni.org/en/article/the-false-promises-of-the-bioeconomy-str ategies

Todd Beer, C. (2022) 'Systems change not climate change': Support for a radical shift away from capitalism at mainstream U.S. climate change protest events, *The Sociological Quarterly* 63(1): 175–198.

Tokar, B. (2013) Movements for climate justice, in M. Dietz (ed) *Handbook of the Climate Movement*, London: Routledge, www.social-ecology.org/wp/ wp-content/uploads/2012/12/Tokar-Climate-Justice-2013.pdf

Tolvik (2017) *UK Residual Waste: 2030 Market Review*, London: ESA, www.tolvik.com/published-reports/view/uk-residual-waste-2030-mar ket-review/

Torgersen, H. and Bogner, A. (2005) Austria's agri-biotechnology regulation: Political consensus despite divergent concepts of precaution, *Science & Public Policy* 32(4): 277–284.

Torgersen, H. and Seifert, F. (2000) Austria: Precautionary blockage of agricultural biotechnology, *Journal of Risk Research* 3(3): 209–217.

TUC (2014) *The Economic Benefits of Carbon Capture and Storage in the UK*, London: Trades Union Congress (TUC), www.tuc.org.uk/publications/ economic-benefits-carbon-capture-and-storage-uk

TUC (2020) *Voice and Place: How to Plan Fair and Successful Paths to Net Zero Emissions*, London: Trades Union Congress (TUC), www.tuc.org.uk/sites/ default/files/2020-08/Just%20Transition%20final_Contents_Updated_ MN%20%281%29.pdf

TUED (2015) *The Hard Facts about Coal: Why Trade Unions Should Re-evaluate their Support for Carbon Capture and Storage*, New York: Trade Unions for Energy Democracy (TUED), in cooperation with the Rosa Luxemburg Stiftung, http://unionsforenergydemocracy.org/resources/tued-publicati ons/tued-working-paper-5-the-hard-facts-about-coal-landing/

TUED (2018) *Trade Unions and Just Transition: The Search for a Transformative Politics*, New York: Trade Unions for Energy Democracy (TUED), in cooperation with the Rosa Luxemburg Stiftung.

TUED (2019) *Bernie's Green New Deal Stands Out, and Now Labor Must Step Up*, New York: Trade Unions for Energy Democracy (TUED), in cooperation with the Rosa Luxemburg Stiftung.

TUED (2020) *Transition in Trouble? The Rise and Fall of 'Community Energy' in Europe*, New York: Trade Unions for Energy Democracy (TUED), in cooperation with the Rosa Luxemburg Stiftung, https://unionsforenergyde mocracy.org/resources/tued-working-papers/tued-working-paper-13/

TUED (2022) John Treat, *True Colors: What Role Can Hydrogen Play in the Transition to a Low-Carbon Future?* New York: Trade Unions for Energy Democracy (TUED), in cooperation with the Rosa Luxemburg Stiftung, https://rosalux.nyc/trade-unions-for-energy-democracy-working-paper- 15/, https://rosalux.nyc/wp-content/uploads/2022/04/TUED_W P15-c.pdf

Turmes, C. (2008) Report on the proposal for a directive of the European Parliament and of the Council on the promotion of the use of energy from renewable sources, COM(2008)0019 – C6-0046/2008 – 2008/ 0016(COD). European Parliament, Committee on Industry, Research and Energy. Rapporteur: Claude Turmes.

UBB (2020) *The 4Rs: Reduce, Reuse, Recycle and Recover with the Gloucestershire Energy from Waste Facility*, 1 December, Urbaser Balfour Beatty (UBB), www.ubbgloucestershire.co.uk/news/2020/12/1/the-4rs-reduce-reuse- recycle-and-recover-with-the-gloucestershire-energy-from-waste-facility

UK Fires (2019) *Absolute Zero*, www.ukfires.org/wp-content/uploads/2019/ 11/Absolute-Zero-online.pdf

UK Government Chief Scientific Adviser (2016) *From Waste to Resource Productivity*, London: Government Office for Science.

UKWIN (2010) *Pyrolysis and Gasification Briefing*, UK Without Incineration Network (UKWIN), http://ukwin.org.uk/knowledge-bank/other-thermal-treatments/pyrolysis-and-gasification

UKWIN (2014) Written evidence submitted by UKWIN to DEFRA, WME 0025, http://data.parliament.uk/writtenevidence/committeeevidence.svc/evidencedocument/environment-food-and-rural-affairs-committee/waste-management/written/9294.pdf

UKWIN (2015) Incinerator refused at Lock Street appeal, *UK Without Incineration Network* (UKWIN), 6 August, http://ukwin.org.uk/2015/08/06/incinerator-refused-at-lock-street-appeal/#

UKWIN (2016a) *Gasification Failures in the UK: Bankruptcies and Abandonment*, www.ukwin.org.uk/files/pdf/UKWIN_Gasification_Failures_Briefing.pdf.

UKWIN (2016b) *Everything Goes Somewhere*, http://ukwin.org.uk/resources/zero-waste-pamphlet/

UKWIN (2018a) *Response to Congestion, Capacity, Carbon: Priorities for National Infrastructure: Consultation on a National Infrastructure Assessment.*

UKWIN (2018b) Josh Dowen, *Evaluation of the Climate Change Impacts of Waste Incineration in the United Kingdom*, https://ukwin.org.uk/files/pdf/UKWIN-2018-Incineration-Climate-Change-Report.pdf

UKWIN (2020) Anti-incineration campaigners praised at February 2020 Parliamentary debate, *UK Without Incineration Network* (UKWIN), 21 February, https://ukwin.org.uk/2020/02/12/anti-incineration-campaigners-praised-at-february-2020-parliamentary-debate, based on https://hansard.parliament.uk/Commons/2020-02-11/debates/D1799344-3D26-4DF0-94C1-3AEB397AF375/WasteIncinerationFacilities

UKWIN (2021) January 2021 debate on incineration and recycling rates, *UK Without Incineration Network* (UKWIN), 27 January, https://ukwin.org.uk/2021/01/27/january-2021-debate-on-incineration-and-recycling-rates

Unearthed (2020) Waste incinerators are three times more likely to be built in poorer areas, *Unearthed*, https://unearthed.greenpeace.org/2020/07/31/waste-incinerators-deprivation-map-recycling/

UNFCCC (2012) Paris Agreement, https://unfccc.int/process-and-meetings/the-paris-agreement/the-paris-agreement

Unite (2019) *Manufacturing Matters: A Political and Industrial Strategy Published by the Unite Manufacturing Combine*, London: Unite the Union.

Unite (2021) *A Plan for Jobs in UK Manufacturing*, London: Unite the Union.

US HR GND (2019) Ocasio-Cortez, A., Tlaib, Rachida, Hastings, Carolyn Maloney Serrano, Espaillat Vargas, et al. *Recognizing the Duty of the Federal Government to Create a Green New Deal*, House Resolution 109, Washington DC, www.congress.gov/bill/116th-congress/house-resolution/109/text

van der Ploeg, J.D. (2009) *The New Peasantries: Struggles for Autonomy and Sustainability in an Era of Empire and Globalization*, London: Routledge.

van Groenigen, K.J., Qi, X., Osenberg, C.W., Luo, Y. and Hungate, B.A. (2014) Faster decomposition under increased atmospheric CO_2 limits soil carbon storage, *Science* 344: 508.

Vanloqueren, G. and Baret, P.V. (2009) How agricultural research systems shape a technological regime that develops genetic engineering but locks out agroecological innovations, *Research Policy* 38: 971–983.

Varoufakis, Y. and Adler, D. (2020) The EU's green deal is a colossal exercise in greenwashing, *The Guardian*, 7 February, www.theguardian.com/ commentisfree/2020/feb/07/eu-green-deal-greenwash-ursula-von-der-leyen-climate

von der Leyen, U. (2019) Press remarks by President von der Leyen on the occasion of the adoption of the European Green Deal Communication, 11 December.

Wall, D. (2014) *The Commons in History: Culture, Conflict and Ecology*. Cambridge, MA: MIT.

Wallace-Wells, D. (2019) *The Uninhabitable Earth: A Story of the Future*, London: Penguin.

Webb, J. (2019) New lamps for old: Financialised governance of cities and clean energy, *Journal of Cultural Economy* 12(4): 286–298.

Weinberg A.M. (1966) Can technology replace social engineering?, *Bulletin of the Atomic Scientists* 22(10): 4–8.

Weinberg A.M. (1967) *Reflections on Big Science*. Cambridge: MIT Press.

Weinberg, A.M. (1994) *The First Nuclear Era: The Life and Times of a Technological Fixer*. Woodbury: American Institute of Physics.

Welsh Government (2012) *Collections, Infrastructure and Markets Sector Plan, Towards Zero Waste, One Wales: One Planet*, WG15009.

Wezel, A., Bellon, S., Doré, T., Francis, C., Vallod, D. and David, C. (2009) Agroecology as a science, a movement and a practice: A review, *Agronomic Sustainable Development* 29: 503–515, www.agroeco.org/socla/pdfs/wezel-agroecology.pdf

White House (2021) *The American Jobs Plan*, www.whitehouse.gov/ briefing-room/statements-releases/2021/03/31/fact-sheet-the-ameri can-jobs-plan/

Wilson, C. (2021) COP26: 'Green' tenement plan could cut fuel bills by 80%, *Herald Scotland*, 10 November, www.heraldscotland.com/news/19705 494.cop26-green-tenement-plan-cut-fuel-bills-80/

Woodin, M. and Lucas, C. (2004) *Green Alternatives to Globalisation*, London: Pluto.

WoW (2020) *Global Green New Deal*, London: War on Want (WoW), https:// waronwant.org/our-work/global-green-new-deal

WoW (2021) Why the world needs a Global Green New Deal. London: War on Want, https://waronwant.org/news-analysis/why-world-needs-glo bal-green-new-deal

WRM (2007) *We Want Food Sovereignty Not Biofuels: Open Letter to the European Parliament, the European Commission, the Governments and Citizens of the European Union*, Montevideo: World Rainforest Movement (WRM), www.wrm.org.uy/subjects/agrofuels/EU_declaration.html

WSP (2013) *Review of State-of-the-art Waste-to-energy Technologies*, London: WSP Environmental Limited, www.rainforest-rescue.org/updates/501/we-want-food-sovereignty-not-biofuels.

XR (2021) *XR Fundamentals: Go Beyond Politics*, Extinction Rebellion, https://rebellion.global/blog/2021/01/05/citizens-assembly-climate-change/

XR MAPA (2022) *Debt for Climate: A Call for Solidarity from Most Affected Peoples and Areas (MAPA) of Extinction Rebellion*, https://rebellion.global/blog/2022/05/12/debt-for-climate-xr-mapa/

XRZW (2020) *Open Letter on Transitioning to a Circular Economy Without More EfW Incineration*, Extinction Rebellion Zero Waste, www.xrzerowaste.uk/view-the-letter

XRZW (2020a) Stop Edmonton Incinerator joint letter, https://stop-edmonton-incinerator.org/extinction-rebellion-letter/

XRZW (2020b) Time To Tell the Truth about Incineration, https://stop-edmonton-incinerator.org/wp-content/uploads/2020/05/2020-05-26-XR-rebuttal-of-NLWA-claims.pdf

XRZW (2021) *Ten Action Points for Camden Council: How to Reduce Residual Waste by 65% and Achieve 70% Recycling by 2030*, www.xrzerowaste.uk/actions/camden

Index

Note: References to figures and photographs appear in *italic* type; those in **bold** type refer to tables.

www.ingramcontent.com/pod-product-compliance
Lightning Source LLC
Chambersburg PA
CBHW070927030426
42336CB00014BA/2572